Tao of Chaos

Merging East and West

Katya Walter

Tao of Chaos

Merging East and West

Kairos Center

1994

Kairos Center

Box 26675
Austin, Texas 78755-0675
Phone: 1-800-624-4697
FAX: 512-453-8378

⚉ Colophon is a trademark of the Kairos Center

Cover design by Kairos Center
Cover fractal by Art Matrix, Box 880, Ithaca, New York 14851
Manufactured in the United States of America

Published in 1994

Library of Congress Cataloging-in-Publication Data
Walter, Katya C.
The Tao of Chaos : merging east and west / Katya Walter, Ph.D.
288 p.   22.5 cm.   "A Kairos Center Book"
Includes annotated table of contents, bibliography,
& illustrations

1. Genetic code—DNA-RNA structure, amino acids
2. Science and Religion—interface, Taoism, spirituality
3. Chaos theory—bifurcation, fractals, analinear dynamics
4. Philosophy—Chinese thought, Plato, Taoism
5. Psychology—Jung, archetypes, complexes, ego, dreams
6. Mysticism—I Ching, synchronicity systems, dreams, chakras
7. Title

QH437.W64 1994   575.1 — dc 20   93-80811
                                            CIP
ISBN 1-884178-16-2 HBK
ISBN 1-884178-17-0 PBK

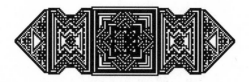

Contents

Preview

The precursor to the *Tao of Chaos* was a German volume called *Chaosforschung, I Ging und Genetischer Code,* published by Diederichs Verlag in Munich in 1992. It created a stir there and was reviewed in newspapers, magazines and electronic media. But in the United States, I couldn't even find a publisher.

So for an American audience, I decided that I must discard that old heavier format with its more Teutonic and technical organization. I developed a new book that was much shorter and bolder, more juicy with personal insights and full of illustrations, and more keyed to the American psyche. This book still holds true, nevertheless, to the original premise that the body and soul become one at the deepest level of structure. The *Tao of Chaos* will take you into that level, into the wonder of an amazing co-chaos paradigm where science and mysticism merge.

As an invitation to your journey, consider these remarks about its ancestor tome that were made by Claus Claussen in the German magazine *Neues Denken und Handeln: Esotera,* November 1992. Let it be noted also that Claussen has kindly given permission to adapt the next-to-last paragraph slightly to fit the format of this new approach.

Universal Life Pattern could be a subtitle for this lofty theme that will pique your interest in the Orient. It might also be called *Breaking a Universal Code,* because it opens the door on a fascinating view of life. Number, more exactly, archetypal number, is the key to this research on chaos theory, Chinese philosophy, and DNA.

Katya Walter, prominent philosopher from Texas, a Ph.D. who also has studied at the Jung Institute in Zurich and taught for a year at Jinan University of Canton, goes to the source of life's dynamic pattern in her book. She describes how the DNA spiral of our linear-minded Western science relates to the analog-style thinking of the old I Ching. She shows that the genetic code and I Ching function through the same chaos patterns, and that the physical system of DNA can be translated mathematically into the psychic system of the I Ching.

Other scientists, and especially Martin Schönberger (1973) in his book *Verborgener Schlüssel zum Leben—Weltformel I Ging im genetischen Code,* have earlier pointed out an astonishing correspondence between the genetic code and the I Ching.

Walter makes reference to this work, but adds a new analog perspective, even enlightenment beyond Schönberger's book, going deeper and wider. Very concretely and beyond speculation, she lays bare a decodable correlation between amino acids and hexagrams. She shows that biochemical laws and old wisdom are connected through this mathematical pattern. It garbs old Eastern truth in new Western clothing. This chaos supersystem is provable with new terminology and computer graphics.

Threading through the awesome labyrinth of this stunning theme, your guide Katya Walter continually startles you back into clarity with her personal engagement in the search for truth. She gives sidelong glances into her dreams, talks of her experiences and frustrations, and even jokes along the path. At such times her tone, normally scientific and yet crisp with a refreshing simplicity, takes on a more poetic lilt.

The author takes an informative stroll through the chaos garden as she explores its profound central theme, approaching it from three distinct vistas: I Ching, chaos theory, and genetic code. This sight-seeing tour is designed to render each path fascinating yet familiar. Otherwise the waves of scientific proof could become too big.

Above all, this carefully crafted work is a treasure trove chock full of jewels. Finally, there is a special paradoxical treasure at the bottom of the chest: without ever leaving the groundwork of science, it moves beyond logic into universal values.

Thanks

This book took a long time to write, and I have many people to thank. Thanks to Marie-Louise von Franz for first noting the odd structural similarity between the I Ching and genetic code. Thanks to Rainer Rieder for friendship along with biochemistry knowledge. Thanks to physicist Greg Romine for the many Wednesday evening talks. Thanks to Röbi Epprecht for the wave frequency talks. Thanks to Rick Goldgar for building an I Ching font that worked. Thanks to theologian John Yungblut for reading with a harmonizing heart.

Thanks to Dirk Evers for warm encouragement along the way, to Ruth Ammann for her feminine wisdom, to Theo Abt for bringing out my strength. Thanks to Andre and Therese Studer, Patrizia Lanz, George Schell, Penelope Yungblut, John Warden, Evelyn Weber, for moral support. Thanks to Silja Gassmann and Jürg Conzett for knowing synchronicity when they saw it.

Special thanks to dear old friend, Zhang Luanling, for tutoring me in the I Ching. Thanks to Tan Shi-lin for discussions on Chinese culture and literature. Thanks to Jinan University for the enjoyable and fruitful year I spent teaching there. Thanks to the many Chinese friends who taught me about China from the inside out.

Thanks to Helen Bonner for her editing skills. Thanks to Marie Adams for Lady, who gave me such loyalty and companionship at the desk during the American rewrite. Thanks to Carol Anthony for publishing help. Thanks to Roberta Hill for her graphics help and taste. Thanks to Terry Sherrell for pro help in various production logistics. Special thanks to the city of Austin for the many ways it has supported me and my work with money and spirit. It recalls the odd fact that *Austin* sounds like *Eastern* in German.

Thanks to John for the many rereads of this book. Thanks to Diana and David for demonstrating from childhood up that both yin and yang are in each of us. Thanks to my father whose honesty brings its own blessing, and to my mother for helping me befriend the dark.

Preface

As I sit down to redo this book for an American audience, what rises before me is last night's dream: I'm in a broad and beautiful land among many trees. It's night. I look up at a huge old tree that's dark against the starry sky in its detail of twig and branch. There is room enough here for all of us, I realize, here in this big, intricately textured park. But I see that some want to cut down the trees and level it all out, so huge throngs of people can gather to gaze up at the sun's glare. I watch dark twigs fingering the remote, untouchable stars. A voice speaks: "Don't turn this into a Copernicus Garden."

Waking up, I remember that I went to sleep wondering how to put this book together. And I take "Copernicus Garden" to mean a parking lot vista where masses gather to honor the bright sun of traditional science with its old rules as the center of the universe.

So I will not cut down the trees and level this book out. It is between you and me, a conversation as we stroll along in a moonlit fractal garden past webby connections of thought that merge to patterned insight. Here hidden delights nestle in scaling patterns of self-similar but never quite repeating beauty. Here the tree of life holds stars in its branches. No matter how huge, this garden stays human-sized because we have a place in it, you and I. No need to cut down the connective forest and level things out for that bright Sol of left-brain logic whose daytime dazzle—so close and glaring—can blind us to the myriad constellations beyond.

Here in the fractal garden, there's room enough for our dreams and dark doubts, our discoveries and slow evolution. There's room to grow in, so we needn't be perfect, or failing that, be cast out. Perfection is end-stopped, but this garden allows change. Here we can still walk with divine nature that is visible everywhere, yet finally unknowable. Its majesty stretches beyond our human ken into darkness, yet it willingly shares as much as we can bear to see.

Katya Walter

Section 1

Pattern in Chaos Around Us

To see a World in a Grain of Sand
And a Heaven in a Wild Flower
Hold Infinity in the palm of your hand
And Eternity in an hour . . . .

William Blake
Auguries of Innocence, 1808

Beyond the Linear Limits

Late one Saturday afternoon in Austin, Texas, I faced a showdown between my left and right brains. Shoot-out time. It was June 8, 1985, and it happened at the East-West Center when Diana Latham urged me to try the I Ching, just once. So reluctantly, I did . . . with a lackadaisical and dubious query. But the response bowled me over! It was devastatingly appropriate—and poetic. Sheer chance, of course.

But when I tried it again later out of curiosity, it worked again. Moreover, the old Chinese oracle's answer again touched some deep chord in me that logic alone didn't reach. A response to wisdom, a welcoming hosanna of "Yes, that's it!"

I was confronted with the notion that this absurd old oracle might actually work, odd as that seemed to my logical mind. Irrational. So strange in fact that I ignored it for awhile. I could not admit the possibility . . . so it chewed just underneath. After all, I was modern, savvy, educated past superstition. A Ph.D. teaching at the University of Texas. Wasn't I? Logic shot the I Ching down. Didn't it?

But I couldn't quite dismiss it. How in heaven's name could that abstract and ancient oracle mesh in such an amazing way with the events in my own modernday life? Rationality said, "Impossible! Chance! Gullibility!"

But the simple fact remained: it had told me apt wisdom, pointed and calm, like a grandparent whispering in my ear. So I decided to explore in rational terms if such a thing might possibly be.

The first trouble was just in finding a handle on it. How could I gather the I Ching by some grasp that my logic-oriented mind could manage to hold, rotate about, and examine in a scientific way? How could I scrutinize the unseen?

Okay, give it the empirical testing that science demands, I said. Don't believe it—check it out. I tried this for several weeks. Each morning I asked, "How will today be?" Each evening I compared the morning's prediction against what had actually happened. There seemed to be a positive correlation between the tenor of the day and the hexagram answer. But maybe I'd created it out of expectation.

So I reversed the order and asked the question after the fact: "What happened today?" And each night the I Ching opened disturbing new insights on my daily hubbub of events—suggesting there was an underlying pattern and connectivity—until I began to suppose that I was amazingly suggestible and susceptible to this old fraud of an oracle. Even though my friends called me hardheaded, smart, and skeptical. Dubious, show-me, and secular.

Either that or it really worked.

So I kept careful records for several months, doing the I Ching each night for the next day's "psyche forecast," much like tuning in the weather forecast, only this time I was getting a preview of my internal weather.

And even through that forest of archaic Chinese imagery, I could see that it probably worked. But how?

Skeptics, especially those of a logical bent—and I used to be one—usually try the I Ching once or twice (if they even deign that much time) and then they say, "Oh, it's just coincidence that this old Chinese oracle happens to tally with my inner reality so nicely this once . . . or twice. Besides, you can interpret anything you want to out of a language and culture so archaic, so far away."

Then they put it down again, partly out of rational scorn and partly out of something deeper that the scorn covers—perhaps it is a fear, some nameless dread of superstition and witchcraft and ignorant acquiescence to an enslavement from the unknown. At least I believe that was the original case with me.

And I am not alone in such a fear. I remember the day when a psychiatrist-turned-Jungian analyst told me that he'd fooled around with the I Ching during his pre-med training at Stanford. He mocked the I Ching to his friends. So they suggested he try it out.

His question to the oracle went something like this: "I'd be a fool to ask you a question, wouldn't I?"

He said, "The answer scared me so badly that I dropped the I Ching as if it had burned me. I stayed away from it for ten years."

"What answer did you get?"

"Hexagram 4—*The Young Fool*. I read it, and while reading, I felt like one . . . my face burning, a young fool chastised . . . and also I was afraid of what might be pointing that out to me. So I called it superstition," he laughed, "which is a different thing, I found out years later, from what actually occurs with the I Ching."

"What do you mean?"

"Superstition is rigid little rituals frozen around symbols. Black cats are evil, so you don't let one walk in front of you. A four-leaf clover is lucky, so pick it. Spilling salt is unlucky, so throw some grains over your left shoulder. Blue wards off the evil eye, so put blue on the baby. That sort of thing. That pat approach will reify your significant moment into a symbol and turn it lifeless. It will record a snatch of bird song even as it kills the singer, stuffs it, sets it into a niche . . . worships it as the symbol of joy while playing that same old bird tape forevermore. 'The stuffed bluebird of happiness. . . '" he intoned in a goofy voice.

"And the I Ching doesn't do that?"

"No. Its analogy is an image-provoker for your mind to dialog with. You project it uniquely into your own situation. So the symbol stays alive. It flows and shifts, depending on what you need to intuit at a given time. The process is subjective, so the I Ching answer really relates to you. It shows you whatever you need to see."

"But how? Why?" Even then I was interested in the how and why. I already knew that it worked, and I was testing my secret surmises and conclusions against this psychiatrist's. But I didn't feel ready to declare anything openly. Not yet.

"I don't know. Beats me. Synchronicity, Jung called it. An *acausal* connecting principle. But it works. Oh yes, even in that funny archaic language, its allusive wisdom is beyond chance, superstition. Jung realized it. And I found out eventually." He smiled. "I'm no longer such a young fool."

Interesting though, isn't it, that it took ten years for the message to soak in past that psychiatrist's conscious resistance. He eventually came back to the I Ching. Sometimes it takes a long time, because the

I Ching is so gentle, so silent, and so abstract. It does not flay you or excommunicate you or shun you. It is not wreathed in lightning or neon or even in flesh.

But I have slowly learned that the I Ching reveals the pattern. Not the specifics of an event, but its underlying pattern. It works through the dynamics of chaos theory, which can predict a trend without specifying its exact details. Discovering this huge hidden intelligence that rests deep in the weave of nature, even learning to communicate with it, can be disconcerting, frightening . . . until it becomes wonderful.

The discovery reveals a deeper truth beyond the limits of what we call normal reality. It exhibits an underlying coherent pattern in the dynamic chaos of nature itself. More eerily, it exhibits a tappable caring that's nestled in the very fabric of spacetime-mattergy. This huge pattern knits the cosmos together in physics and metaphysics. It unites the objective and subjective, the quantitative and qualitative, the alpha and omega. Its vast dynamic shapes us, body and soul.

But even now, despite extensive experience, I still understand why people fear such a possibility lurking beyond the sensory reality. What is this strange domain of power? It sounds rather like superstitious enslavement to some bogey of the imagination . . . and that is worthy of fear.

Fear is a reasonable response. It sees the psychic loss in giving oneself up to superstition and voodoo rites, in relinquishing personal power to brutal gods with canny, greedy priests, to wizards, healers, or gurus only too willing to manipulate their followers if given a chance. It fears an abandonment into destructive and orgiastic fantasy, a letting-go of choices, a precipitous lapsing away from logic to become lost in the enchantment of the deep. The descent looks too dangerous into the primal power of raw archetypes. It seems the epitome of romantic release to Dionysius, where one fears and yet follows a blind pull into the passionate unknown.

In here is an emotional powerhouse. In here, people "lose control" and argue or rape or fall blindly in love or seek the grail. They kill others in holy war or devote the rest of a short lifetime to the cure of AIDS or become inspired to paint a masterpiece. Moments or years later, a person can say bemusedly, "I don't know what came over me," or "Something possessed me," or "I felt driven!" or "The devil made me do it," or "I followed my bliss and it has blessed me."

This mysterious domain is not logical. It has its own reasons that reason does not know. To quote a bewildered Woody Allen, "The heart wants what it wants." But a passionate vision can so easily lead to bitter disillusion, to mob violence, to some idiosyncratic utopia where a demagogue—religious or social or political—holds sway and demands blind faith of spellbound followers who do battle for the cause. Some cause. It can idealize a saint or a demagogue to follow, fix upon some vile enemy to conquer, so that half of its huge polarized energy is projected away into the outer world to become wooed or defeated, driving us without free will.

Yet there is also a fear of *no* passion. Cool Apollonian logic can survey this universe with a despairing sense that it is caught in some limitless, endless, mindless design that also leaves no room for free will. Sartre showed us the architecture of this existential prison with no exit from a godless hell of cold linear logic. It locks us into a gray, chilly basement whose flat expanse of pointless concrete data cannot conceal the futile bones buried beneath. Here is dead-end living.

Both extremes—passionate romantic or cool logician—reveal a paucity of perspective in the West for more than 2,500 years. It has split us into romantic versus classic, liberal versus conservative, left versus right, heart versus head. But it is possible to encompass both poles within a larger, transcendental third stance. This paradigm is cradled in chaos theory, that amazing new science of the 20th century. It reveals the I Ching to be a model of chaos patterning in microcosm. It even connects science to spirit.

The I Ching is a spiritual tool. But not a religion. Religion, after all, is a cultural structure built to house the human spirit. Its doughty edifice is not exactly spiritual. Even the most moderate, gentle of religions is systematic, socialized dogma. Frequently it is codified by the few for the many. Often it is based on some divine revelation that is declared inaccessible to the rest of us. We should just take their holy word for it, even if we have doubts.

I always have doubts about something that won't let me doubt. I see flaws in every belief system that we share in society—even those I embrace, like democracy and literacy and ethical responsibility. I myself am not a perfected system—never will be, since my own life is in continual evolution—so why should any other system be perfect? I look around me and don't see perfection. But I do see change. Evolving ways to perpetrate vices and honor virtues.

All this said, it is no wonder that so many modern people are skeptical and fearful of this realm beyond logic, dubious not only about religion, but about spirituality itself. After all, the banner of religious fervor has incited holy wars, ethnic cleansing, first-born butchering. Practically every culture has sacrificed itself or another to appease some bloodlusting image of religious righteousness.

Most religions no longer kill to prove their dogma's strength. But even as the image of god has evolved and gentled, it has also become so bland and generalized and ineffectual as to be quite dismissible by many. Existentialism, communism, fascism, nazism, capitalism, scientism—these are the new religions whose secular gods have offered current dogmas far from any divine source of power. Many people currently believe in some secular -*ism* rather than a religious -*ism*. Yet these too fail, sometimes horribly.

So god has died . . . god with a little g, the generating, organizing dynamic that permeates the universe as ubiquitously as atoms and energy, as implicit as time and space. Not really died, of course, but the image of god has died within so many of us, and this loss has been projected outward onto our universe as the realistic view.

God? Dead and gone! Good riddance too, many have said. Why believe (during such a modern, technological era) in that primitive superstition called god? Many have tacitly agreed with Karl Marx's assessment that religion is just an opiate of the people.

In fact, as religion faded in the society, drugs began to promise the release and ecstasy that spiritual faith once did. People climbed an alcoholic or acid or crack or heroin "Stairway to Heaven," and in a strange reversal, opiates became the release-cult of the masses. We became addicts to escape this nitty-gritty, drab world without lofty spirit that we'd finally engineered for ourselves.

But unlike the spiritual bridge to another realm that's found in religion and mysticism, habituating drugs can only offer a road away from, not to. This escape into addiction is triggered by our search for psychic release from the data-mad mire of ordinary life. For the beleaguered animal spirits, it promises a momentary release, but no uplift into transcendent spirit. Addiction finally becomes just another dead end, not a path to enlightenment.

At some deep level, we already know that in the modern war on drugs, we have met the enemy and it is us. The enemy marauds within. It is our own estranged hunger for spiritual meaning in this modern, linear-blindered world. We starve our ineffable soul and surfeit the flesh. Someone once told me wryly that *ineffable* is just a Latin word for unfuckable, and *psyche* is just the Greek for that four-letter Anglo-Saxon word called soul. Our culture has suffered an atrophy of the soul. The modern psyche hungers for meaning even as it is dying of self-inflicted starvation. Carl Jung diagnosed it in his book, *Modern Man in Search of a Soul.*

Where can we turn for our soul's nourishment? What can succor its emaciated yearning? When we gaze around, starved, at the havoc wrought by a world of fanatic *-isms,* it is no wonder that so many are disenchanted, leery of belief and this crazy gullibility that we humans exhibit whenever power radiates from the analog dark, from that realm of passionate connection beyond logic. Succor, indeed.

But soul-starved, must we go ahead and swallow blindly the old rites and ways? Are we so desperate to regain spiritual connection that we must accept its past corruption along with the vigor? Where do we find an answer that will not traduce us, betray others, again straight-jacket the image of god into some very limited, culture-bound vision? I cannot point out to you a single religion or *-ism* or creed that is beyond human distortion.

And yet with the I Ching, that most solitary and meditative of oracles, for me there is something finally impervious to human meddling. The subjective stance that I take toward it is already a given. I can stand in its answer and look around my life to reach past my own frailty for some core of truth beyond the projective screen of my own psyche and mere chance. My experience, recorded over enough time, has shown me that this is so. But how can it be?

At first I couldn't believe it. Not logically. The peculiar beauty of its poetry intrigued me, but the accountability of its answers fascinated me even more. Keeping a record became my introduction to the realm that goes beyond mere Cartesian logic, into the wordless way of the Tao. I discovered that in here is logic *plus* analogs.

How does a mere I Ching oracle answer questions? Much like a computer program. It uses a mathematical algorithm to relay its response in an analogy from the archaic Chinese past. (A quick and accurate method is given in Section 4, "Consulting Procedure.") This analogy—or verbal analog—really does correlate in a fit far beyond chance. Perhaps I first realized it so quickly because of my love for poetry. At any rate, soon I saw that the images of the I Ching are not just poetry. Your hexagram answer will diagnose the situation in a specific verbal analogy that faithfully conveys the dynamic pattern of that realtime event. Its analogy offers advice on how to cope with the situation—in other words, how to go through it in Tao.

Its right-on accuracy is easiest to recognize in issues about your own past. A question on some current issue can also bring that shock of recognition. It is the query about the future, though, that gives the most room for resistance and doubt. The answer can seem irrelevant, improbable, absurd—so, of course, your mind rejects it and thereby dismisses the I Ching too. An eerie accuracy will appear, however, after the fact, when the issue is resolved, so that by keeping and re-reading your daily log to compare answer with actual outcome in an empirical way, you are likely to discover that the I Ching actually saw around a corner of the future in a way that your ego could not.

With growing excitement, I saw that the I Ching actually works; further, that its accuracy might have a basis in the new science of patterned chaos. It led me to five years of research in Zurich, a year in China, and nine years of writing. It has been worth it.

I also think that your I Ching experience, if honestly recorded and checked and mulled over time, will show you a correlation so far beyond chance that it will open a whole new realm of connective pattern wheeling at the edge of perception. Keeping your record is very important because things at the liminal edge—dreams, fleeting fantasies, synchronicities flashing their subliminal clues—these are easily dismissed or distorted by the controlling ego as they pass into its bailiwick from beyond. Ego denial will send them back again.

Insights arriving from beyond the fringe are intensely personal. They shift in import on each review. They offer a series of continually altering slants on the hologram of your life from the changing perspective of its passing days. Each moment becomes a commentary on what came before and teases your ego with glimpses of what lies ahead. Always the message is relational, always personal, even as it can slowly reveal a universal heart of order and meaning. Discovering this deep order within chaos brings a kind of inner knowing or wisdom. It is the aha! of revelation that cannot be described in mere words, but only felt. It gives meaning to being.

The main difficulty with this holistic realm, of course, is that it flickers at the edge of awareness. It signals from beyond the fringe. It keeps fleeing from direct gaze. Thus it must be approached with the sidelong glance of holistic pattern recognition rather than the pinpoint glare of linear dissection. Much as the peripheral vision can perceive a faint star where the direct gaze cannot, so does the liminal glance of awareness discern a reality that the ego would sooner not notice. The strange face of analog truth can be especially frightening to a rigid, impermeable, brittle Western ego. A logic-bound ego expects only demons beyond its limits. It marks the map of its psyche with warnings: "Beyond these gates there be monsters."

Ego stands at the gate. So it is important to develop a strong, flexible, savvy gate-keeper. That is difficult enough. But turning this ego into a tool of the larger self? . . . that's even harder. Learning how to use ego as a mere tool, and when to put it down, takes work. It includes such trifles as bothering to record your I Ching answer and recheck it later, grabbing and studying your dreams before they slip away, discerning the archetypal framework of your life instead of just letting it flicker by as foolish capering events in a pointless film.

A single question pervades: What is the dynamic behind the dance? The shape of truth beyond the veils of Maya? You pursue an insight glimmering at the edge of a dream, an uncanny coincidence, a Freudian slip. Even a psychic, or Tarot cards, or palm reader.

But it is the I Ching, more than anything else, I believe, that can weave these glimpses into a connective and definitely scrutable pattern. It can show you the shape of your inner life. This design past the grasp of the five senses was named the Tao (the Way or Path) in ancient China. Its truths are so hard to haul into the harsh spotlight of awareness, and harder still to articulate. Laotse opens his *Tao Te Ching* this way: "The Tao that can be spoken is not the Tao."

Yet it is exactly this meaningful pattern in psyche and matter that we will pursue throughout this book. Have you ever tried to explain how to dance, instead of just doing it? Or described a tennis match and perforce left out the hushed thrill of the crowd? Or made love—by phone? Then you know how tough it is to put a dynamic into mere words. No wonder the Tao that can be spoken is not the Tao.

A most peculiar situation. The path through these pages will pass into the nearly inarticulate right brain where holistic timing and spacing give us glimpses of huge coherent design in the I Ching's 64 hexagrams. We will follow that path back into the left brain for a distinct articulation and logical tracking along the DNA double spiral to the 64 codons. This will lead us into the amazing paradigm of co-chaos. Here is an unparalleled check-and-balance supersystem that evolves life, both its physical structure and its mental flow.

How does your life flow? It has its unique pointillist design, graphed in the raw data speckled through your days and nights as events. Sometimes there is evident logic to it, but more often there is not. You inhabit life uniquely, in a never-to-be-replicated zone of the hologram. Whenever you search for connective meaning in all this chaotic welter, you are simply trying to puzzle out the chaos fractals that shape your days.

How to go about mapping this secret design in your own life? Prepare to enter unknown territory. It is almost inaccessible to the bare senses, which gauge and report a thin conscious foam over the deeper current of events. The turbulent fluidity of this deeper level has long been a mystery. Its enigma hides both around and within us, because our sensory organs are not geared to apprehend it through direct perception. The psychic life lies above and below the nerve receptors, and thus its shape has to be pieced together indirectly through the aha! of pattern recognition—quite a task for your insight. No wonder it is jokingly called a sixth sense.

Just because this realm is invisible, though, that doesn't make it unreal. Gravity is real, even though we only see its effects. Imagine a cave man pondering what kept his feet on the ground. Nor can we touch a mental complex with a sense organ, yet psychology has showed that such complexes do exist, has even "proved" them by charting their consistent significant patterns in behavior.

We cannot touch a dream. Yet dreams do exist. We moderns are beginning to discover—rediscover—that there is holistic meaning coded into these wild stories. If the forgotten language is recovered, a dream can yield up insight showing that it is no mere "garbage in, garbage out" effect. It offers wisdom beyond consciousness.

How hard it is, though, to grasp this profound meaning hidden in the secret realm. Many cultures have left it to the mystics and gurus and prophets visited by god. They stand out on the threshold between the known and unknown and proclaim some design that lies far beyond the normal vision. Seers have crossed that threshold in trance, insights, dreams, fantasy, in wordless dervish motion. Sages and mystics have flowed bodilessly into the dark mystery and dragged back a revelation into the light of day.

The I Ching can bring revelation to the rest of us. We ordinary mortals can glimpse pattern beyond sight and taste comfort beyond bliss. How? By discovering that the I Ching is merely a spiritual version of DNA, and both of them are merely subprograms of a deeper, even universal order called complementary chaos. For the scientific mind, this allows the security of an explanation. For the mystical soul, it opens an endless vista on wordless beauty charged with transcendent connection.

Over time, with experience, you can track the I Ching's abstract patterns in your realtime events. Ordinary you, no saint or mystic, can bring the holistic gestalt of your right brain into sync with your left brain's linear analysis. You can open the connecting door wide enough to let your logical mind peek at that vast hologram wheeling in the shadows. You won't see everything in there. You couldn't bear it anyway. You'll just hover on the threshold and flash onto whatever portion of the connective pattern you are ready to discern. For it is overwhelmingly powerful, this vast mute design. It carries the numen of the divine plan. If you fear it too much, dread, doubt, deny it too much, the doors will shut down again in sympathy with your fear. If you hold it in too little awe, though, and leap in smugly past your depth, it will shortly teach you its fathomless power by swamping your mere human ego's puny resources.

The I Ching is a guide into the heart of meaning. It is this power to touch the inmost recesses, rather than any social dogma, or revenge in god's name, or control for gain, that is so striking. It merely answers questions with verbal analogies (based on analog math, as we shall see) taken from a lost civilization and time in ancient China. Yet it is extremely potent. I have seen it shake skeptics with some "uncanny" answer. Mere chance, they frown. Foolish luck. Or perhaps . . . magic? But it is not. One might say that a lightning bolt blasting a tree is magic or luck or chance . . . if one knows nothing about electricity and how it builds up its polarity and discharges.

God has carried lightning and thrown it around as bolts in some religions. Lightning is a riveting natural force that believers can see and worship as divine power. But the I Ching experience is not flashy. It is just me and it. Or you and it. It needs no society or brethren or coven of convoked I Ching worshippers who deify and reify its rituals. The many groups who tried this dogmatic approach grew stagnant and withered themselves.

The I Ching, old as it is, remains fresh. Its peculiar strength lies in its use of the chaos supersystem that is so ancient, yet ever-fresh in application. Long after its first appearance in history (scratched on bones some 5,000 years ago), the I Ching is still being interpreted and re-interpreted in what Nathan Siven calls "the uncountable Chinese commentaries on and studies of the Book of Changes."

The I Ching makes no claim to being objective. Instead it is apt. Holistic. Relational. You ask it a question about your life. Mildly it replies like an old sage with a story from the past, who at the end murmurs, gazing at you pointedly, "The moral of this story is . . . ?"

And you must figure it out from there. Apply it or not. Believe it or not. It's up to you to learn from it. Or not. How different this is from the pushy Western way of instilling order, laws, even ethics with the sword of righteousness.

Westerners who only appreciate "objective" thinking will forget how much truth lies in the holistic, relational, subjective right brain. Objective analysis has been the goal of Western science. This moved into high gear, I suppose, when we adopted Copernicus's idea that the earth is *not* the center of a universe tailored just for us by a Christian god. His upstart notion flouted the religious belief of the time. It was such heresy, in fact, that perhaps Copernicus was lucky to be on his deathbed when his book, *De Revolutionibus,* first came out in 1543. But it fostered a mental revolution that eventually gripped the West: the logical, detached study of an object instead of getting mired into subjective connection with it. Stay objective!

Why did we honor this dictum of objectivity for so long? For the best of reasons . . . it worked so well. At the level of Newtonian re- ality that our senses inhabit, it has proved scientifically valid and reliable. So successful, in fact, that gradually this objective stance took over modern culture, not just science. It championed the goal- focused left brain over the holistic right brain. It reinvented the universe as a cool model of linear logic—for example, by erecting straight-edged buildings that deny emotional relationship, unlike those old Greek temples that applied their sly art of proportion to trigger the visceral response of "That's art!—not just engineering."

An increasingly "objective" society weakened its blood and clan ties, gradually moved into disposable relationships and rotating stepchildren. Lifelong loyalties watered down to a flux of solitary TV dinners, temporary residences, interchangeable acquaintances.

Painting, sculpture, dance, music, drama—all boasted the message of meaningless, empty, frantic life—which itself reinforced more mechanistic separation and alienation. Literature bred its anti-heroes whose great frustration was in finding no exit from this stale horror, no relief from a bleak dis-topia stripped of its subjective frills like ethics, sentiment, and ultimate meaning. Cult and culture heroes committed their media-attended suicides, slow and fast. Eventually the globe itself seemed bent on self-destruction.

Even God died. The scientific mindset found god to be objectively unprovable, and religion merely a projection of the desire for life beyond this frail cage of our skeletons. Many intelligent people forswore universal meaning, since that must finally remain subjective. For a life's meaning is to be found merely in its own experience, in each fleeting moment, scientifically unprovable and nonreplicable. Thus reduced to chilly statistics and lab experiments, the vast population of anonymous urban society lost sight of the vast wonder in each personal unreplicable moment, in our unique multiple visions that when overlaid, can connect the world into a holistic flow of qualitative meaning.

As the brilliant Sun King of logic rose overhead to rule in Western society, the queenly dark constellation of meta-meaning became the crazy aunt in the basement. She was banished to a cultural doghouse far out in Coventry, beyond the pale, way to Hell and gone. But of course, Hell is not buried underground but in the darkly repressed and eruptive unconscious of a proudly logical society. Drugs, crime, violence, pollution—that bloated shadow side of our linear, take-charge culture has kept proving to us again and again that we don't really have things under control. Not at all.

But seeds of change sprouted from underground during the 20th century. Freud opened a royal road to the unconscious through dreams and Freudian slips. Jung tracked archetypal patterns in individual and collective dreams . . . and, of course, dreams can only be experienced or reported subjectively. New "soft sciences" emerged to chart the myriad variations on holistic reality that become evident in a comparison of languages, psyches, cultures.

Even that most respected of the "hard sciences"—physics—faced a strange subjective flow within its own technical data. Einstein's thought experiments revealed that the spacetime fabric can only be measured relatively, and that matter and energy are just disguised versions of each other. Quantum physics proclaimed light to be both particle and wave, and these are somehow not simultaneously and equally measurable. New geometries emerged to challenge Euclid's old theorems, new algebras antiquated the sturdy old rules named for that Arabic chemist Al Geber. The very foundational laws of physics and math became modified or abandoned.

We have discovered with a shock just how limited is our sensory view of matter. Of energy. Space. Time. With new tools, we proudly stride out to chart new realms beyond the naked senses. Artificially extended perception allows us to enter macro- and micro-realities that were formerly invisible. Our new sense extenders take us into places where we find the old rules of Newtonian science just don't apply. Data from beyond the sensory fringe is drawing us into a vast new patterning where the old linear paradigm fails us. Nonlinear dynamics whisper of unimagined meshings. The traditional scientific mindset cannot describe this new universe of holistic relationship that is whispering all around us. In us.

We are scientific giants of the sense organs, with technologically enhanced enormous eyes, ears, noses. Yet our tiny human bodies have not kept pace in psychic growth. We're still pygmies of the soul, forcing it to fit into an old mechanistic model where linear logic still holds sway . . . where a soul cannot be weighed, so it must not exist; where cosmonauts scornfully radio back that their lofting rocket has cracked no heavenly crystal sphere.

But scientific objectivity does not finally exist. It's a fond delusion. Even science nowadays admits that its research is forever limited by the cultural, even the species mindset. What each of us sees and how we make judgments on it must inevitably depend on our varying stances in the hologram of reality.

In science a paradigm shift is taking place, a colossal adjustment that reorients the human mind after 2,500 years of accelerating emphasis on linear logic. For so long now, we have measured out our reality in coffee spoons, inches, kilos; we have linked cut-and-dried units into a cause-and-effect sequence to reach a specific goal. True, it has worked well enough at the daily mezzo level described by Newtonian science.

But suddenly we're finding that we've gotten lost way down in the weave of scientific details, counting out our steps, tracing some narrow thread of linearity in a pacing measured tread. But the web of lacy connectivity below and above Newtonian science offers the wonder of observer-sensitive electrons and relative spacetime. We find here a huge balancing of cosmic constants that apparently even takes human life into account. We discover patterns that stun our ability to grasp, some vast order that holds deep in apparent chaos, as though the cosmos itself is a giant woven design.

This patterned chaos can't be fathomed by logical analysis alone. Instead it is perversely frustrating . . . like life itself. It cycles without exact predictability, much in the same way that you go through your own reiterating cycle daily. Your life's events are patterned similarly yet different each day. The way you find out what's going to happen today—exactly—is by watching it happen. Days cycle again and again, but in subtly new formulations. Every once in a while there's a big shift. Then it settles down again. Unless you're in some really wild and intricate pattern. Then it may take a long time for you to eddy back into something less perturbed.

The constant in all this is just yourself. You act as the focus for all this shifting data as you seek your own life's hidden patterns amidst the apparent chaos. Your heart emphasizes the relational quality of your life's events rather than their discrete quantity. At times, it can look and feel totally chaotic. But patterns do lie hidden there, meaningful patterns. And occasionally you do glimpse them.

Now it is time for a new way of embracing the cosmos, each other, ourselves. The paradigm set forth in this book shows a remarkable co-chaos supersystem. It codes for all life, even including the life of the mind. It generates the DNA that builds your flesh and also the flow of thought that allows you to recognize and respond to the Tao. It bonds your soma to your psyche, merges your matter with your mind. Its universal root nourishes both physics and metaphysics. It molds the limits of your body even as you are stretching to hold its vast pattern in your mind.

Tapping into this huge pattern can give conscious meaning to your daily motion and instill spirit in your flesh. The whole cosmos is created by this generating, organizing dynamic—so for short, let's just elide it into that single blunt old Anglo-Saxon syllable: god.

Welcome to the world of complementary chaos all around us.

WORKSHOP SESSIONS CHECKLIST

At the end of each chapter is a short section for follow-up discussion and workshop activities. Some topics are philosophic, while others are more experiential. The intention is to activate both linear and holistic approaches to the book material. Of course, all topics may be modified or used as an impetus to tailor your own specific plan.

Use shared leadership. Ask for volunteer teams of two to facilitate each of the 12 workshop sessions. In Session 1, discuss the book's table of contents and invite pairs of volunteers to sign up for each of the remaining 11 sessions now. Use a chapter for each session, except put all of Section 4 into one session. A final 13th session may also allow a windup party. Invite Eastern and Western dress and food.

Use both circle and row groupings in the activities. Be sure to mix experiential and abstract exercises. Utilize each member's special experiences, travels, and expertise. Allow enough leeway in your topics and approach so that any member can feel comfortable in leading the group at least once according to his or her own personality style. Allow for both traditional and innovative approaches. Pace the quieter moments such as meditation, slide viewing, and hearing a quotation or poem with more active events such as small group discussion, role-playing, chanting, dancing or finger dance. (Some of the suggested activities may seem silly at first glance, but they have their purpose.) Use music, video clips, movement, pacing, volunteer comments to keep the members engaged and lively.

Allow for the unexpected. Events tend to time out a bit differently from what you anticipated. Let laughter and goodwill ease any bobbles and surprises. But arrive on time and stick to the session time limit as much as possible, so that people can plan accordingly.

Open and close each session with a brief ritual to establish group identity. *Suggested opening:* use the ringing of a bell and the lighting of a candle to help everyone "be here now." Then repeat after the leader: "Lines of power, circles of concern. We open our hearts and minds. We hear truth spoken and unspoken. We honor the vast universe around and in us." *Suggested ending:* join hands with those next to you and say, "Life is more than we can see. We are more than life can say. We are lines of power, circles of concern, spirals of change. Merry meet and merry part, and merry meet again!"

WORKSHOP 1

This first chapter is a basic introduction to the network of cosmic patterning that exists beyond our ordinary sense perception. Choose a brief quotation from the first chapter to introduce each of the following points . . . or develop your own topics and activities. Set up a timetable to introduce and do each activity. Freely modify and mix cognitive and experiental topics, but be sure to have both.

❣ Do the 20th century sciences with a subjective component have an impact on you personally? Do they change your world view? How so, or why not? Recall a personal instance where science has had an emotional effect on you. Was it positive or negative? Why?

❣ When you say, "I'm being objective about this," what do you really mean by this remark? Do you consider yourself to be more objective, subjective, or balanced in your everyday life? Give an instance. Do your daily activities reinforce a certain bias or habitual mental stance? Are you satisfied with your stance?

❣ What is your view on the existence of god? On the existence of an ultimate, all-encompassing pattern? Are they the same thing? Do you personally experience any evidence of cosmic patterning in life? Think of an instance or example that illustrates your point of view. Would cosmic patterning have a physical component? Mental? Would it involve some sort of universal body? Universal thought? Caring? Explain your viewpoint and listen to others.

❣ Invite the group to approach connective pattern by passing through the threshold of the ordinary senses into the amplifying chamber where psyche intuits meaning. For instance, you might invite people to experience a selection of items that you have gathered to stimulate sensory experience—perhaps a bowl of corn chips to pass and taste, a toy horn that you blow or cymbals that you chime, a rose with thorns to smell and feel, a heavy clump of amethyst crystals, a kaleidoscope to view. If the group is large, hold up each item, and with a few words, evoke its taste, sound, smell, sight, touch. If the group is small enough, arrange people in a circle with a central table containing the items. Introduce each item and pass it around the circle to feel the textures, hear the sounds, smell the flowers, taste the finger food. Ask people to savor each sensation, to pay absorbed and particular attention to the quality of each experience. Read relevant short passages that you choose from the first chapter as the items are being passed from person to person around the group.

❦ Ask the participants to close their eyes and allow a favorite taste, smell, touch, sound or sight from memory to float up into consciousness. Invite them to open their eyes and call out a word or phrase describing these favorite sensory delights at random into the center of the group—for example, "cold sweet watermelon on a hot day," "the smell of popping corn," "my cat Zorro's soft fur."

❦ Invite people to shut their eyes. Play evocative music as you ask them to move back into some personal moment where a specific taste, sound, or smell triggers the recall of a significant event. Invite them to walk in memory over that threshold of sensation into the great room of amplifying emotion and connective detail from the past. Let the memory evoke the event. Walk around its hologram to see what really happened there, how it happened, even why it happened. Consider what this event has meant for your life. (Allow 5-10 minutes to lead people slowly through this meditation.) Close the meditation by bringing people back into the present time and place.

If it is a very large group, invite some volunteers to discuss their triggered memories. In smaller groups (or a large group that has been subdivided), give each person a chance to speak. Recount your remembered event as though it is still vividly before you. Describe its network of detail and emotion. Listen appreciatively to the others describe theirs without offering criticism or advice.

Ask if people see any common traits among the various accounts. Notice that each cohering pattern of experience is triggered by a sensory memory. Linkage exists between sensory detail and the psyche's ability to intuit a larger pattern of meaning.

❦ After the closing ritual, spend 5 minutes on feedback among the group members on what people have especially enjoyed or realized or would like to try in future. Remember, it is your group. You will together fill the workshop's form with the contents of your unique members. Each can influence it by participation, comments, and co-leading.

To Have or to Hold

As I began to study the I Ching, I noticed odd parallels between its ancient system and modernday Western science. The earliest clue came in Marie-Louise von Franz's 1968 essay called "Symbol des Unus Mundus." In it, she remarked that oddly enough, the I Ching and genetic code both show a 64-part structure. The I Ching and the genetic code? Might they really have something in common? I set out to explore this notion. But where to begin?

Well, why not number? Number is logical, right? Maybe I could get a handle on things by analyzing, first, the I Ching number structure, and second, the ancient Chinese mindset that created this extraordinary document marrying abstract number with poetic nuance. So I looked. And gradually I found out that the Ching is not merely linear, logical, and binary as Leibniz had thought. Instead, it has a peculiarly "nonlinear" vision. It is both analog and linear. These merge to create a vast dynamic pattern of complementary chaos—or co-chaos, for short. This is its paradigm.

At the core of this paradigm sits the new science of patterned chaos. In it, our ordinary environment goes through evolving cycles of time and space as energy moves matter into patterns that are self-similar, yet uniquely varying, and they exist all around us on scales large and small.

Number is the root. You will soon find that this book is a paean to number. But it's not a math book. This number is not the cold stuff that we stack into linear sums and statistics. It is warm. It weaves the very breath of life, your flesh and thoughts. Its webby woof and warp interplay to make the texture of your days.

Here you will find number—normally coolly logical—going to the intuitive heart beyond logic. It probes deep qualitative values that we've almost forgotten, focused so long and hard on taking the discrete measure of things, on statistics rather than relationships, on a cool observer's distance rather than the intimate I-thou bond.

This book will not scorn Western science, but rather, enhance it. We'll not discard linearity, but instead pick up something more: the analinear. Analog *plus* linear. Some people call this "nonlinear."

But the word nonlinear is fairly silly. Mathematician Stanislaw Ulam pointed that out frequently. He said calling something nonlinear is like saying most of the animals in a zoo are non-elephants. It doesn't explain much. For just as most of the world's animals are non-elephants, so are most of the world's processes "nonlinear."

The term actually means not *just* linear. It means linear-plus; it combines the linear chunks of cause-and-effect logic with the flowing proportions of analogs to birth a transcendent third condition. For this reason I prefer to use the term *analinear,* combining both modes to create a synergistic third state.

In the West, formal thought has mostly concentrated on the linear problems, ever since the early Greeks and going on down through 2,500 years of history. Why? For the very good reason that they could be solved. The far more numerous analinear puzzles were just too hard to get a handle on. So people mostly stuck to the fewer strictly linear problems that offered readier solutions.

But analogs abound in daily life. Your psyche vibrates to their hidden quixotic resonances all the time. They imbue events with a force beyond mere linear logic. Networks of connection shimmer and shimmy everywhere, impalpable yet obvious. They shine in commonplace phrases like "That rings my chimes!" "I smell change in the air." "What goes around comes around." "The way you choose 'em is the way you lose 'em." "I got strange vibes off him." "We're on the same wave length." All these slang a*nalog*-ies carry in them the collective wisdom of the folk. They imply networks of relationship. They vibrate with associative meaning.

And what are analogs? Analogs are simple. They are intuitive. They make comparisons. They rely on proportional relationship.

Analog Face   Linear Face

Two Faces of Time

Here are two watches. Both show the same time: 9:30. The watch on the left has an analog face. It reveals the proportions of time in round, related cycles—for example, a child learns that the hour hand takes twelve hours to revolve in its sweep around the dial. But the minute hand traverses those same markers in just an hour. And the second hand does it in only a minute! All this gives the child a sense of the overall proportions between hour, minute, and second. Each measurement relates to the others, with the 60 as the common factor.

The other watch—square—has a linear face, sometimes called digital (even though it's not really—it doesn't count by units of ten like the digits on our hands). This watch chops time into discrete units without showing their proportion to anything. It's just a string of numbers jerking by on the liquid crystal display, and you must know the cultural convention to realize that after 9:59, the hour digit will flip on into 10:00. Otherwise, the minute numbers might just keep getting higher to 9:60, 9:61, 9:62 until they disappear off the edge like a calculator. Or a hundred minutes might make an hour and a hundred hours a day. Why not, logically speaking? After all, it's often called a digital watch, and the 10-base seen on our fingers appears in most places—in our money, the metric system, the Dewey Decimal system, and so on.

You just can't intuit the over-all proportions in our time system by a linear watch, and it's for this very reason that some airline gauges have gone back from linear measurement to analog—the digital readout didn't really convey the proportions of things, but only a raw score. So you've got fifty gallons left in the tank. So what? Is that almost full or almost empty? An analog dial will tell you the vital difference, but a linear dial won't.

The analog principle doesn't apply just to raw numbers either. It is usually fleshed out to a subtler level of presentation. Its best-known child is the verbal analogy—a sentence that compares things. For instance: "Marilyn was to the '50s what Elvis was to the '60s." That remark is weirdly opaque to strict logic. Marilyn who? Elvis what? What did these inadequately identified people do? Where?

But we can make swift analog leaps through history and get it. After all, Marilyn Monroe was a female movie star of the 1950s, Elvis Presley a male rock star of the 1960s. Both sprang from poverty; both rose to pop Olympus as voracious and vulnerable sex symbols; both lived glamorously and died tragically in dramatic circumstances; both became cult figures. It's all there. If you know their history and mystique, you catch the drift of this analogy right away. And we can even set this free-wheeling analogy into math:

Marilyn was to the 1950s what Elvis was to the 1960s.

a is to b as c is to d

$a : b = c : d$

$a/b = c/d$

Suddenly it's algebra! Notice that this verbal analogy-cum-ratio is not just a comparison between 2 things, but actually, among 4 things. Between pairs of pairs. It describes a *relationship between relationships!* What we're discussing here is two different stars in two different decades, and the relationship among these 4 things.

Or take circles. In Euclid's plane geometry, you can make circles of any size, anywhere, as often as you want. But you must always use the same ratio: circumference is to diameter as 22 is to 7.

A pair of pairs. This setup starts the analogs rolling. Here's where the analog domain develops its relational heart. Its parts *must* relate to each other. How different this is from linear number, where each chunky unit stands alone, is enough, signifies just in itself.

The circle will always have the same ratio of circumference to diameter. It's the only way you can fit a circle together. Cultures around the world have honored this ratio. In Western math it is symbolized by the sixteenth letter of the Greek alphabet, pi or π.

And talk about pi in the sky! When you turn this ratio into chunks of linear number, they trail off to infinity! Chaotically! How come? Divide 22 by 7 to get the decimal version. It's 3.141592654 . . . on and on and on. Forever. Literally. There's no end to it. The number just keeps extending; it's been mathematically proved. Also it's been proved that you can't even predict what number is coming up next in this infinity string. There's no pattern. In the words of Paul Davis, "decimal places of π form a completely erratic sequence." Unpredictable. So from the proportions of a simple circle, we've generated endless chaos. The reasonable, *ratio*-nal proportions of 22/7 have become ir-ratio-nal. It becomes a linear string of ever-extending chunks of number.

But the Western cultural mindset became not just linear, but also *binary*. What does this word mean? Well, it sets up lumps in a row to count. But a very short row, with just two lumps on a yes/no shunt. Binary means yes or no, on or off, 1 or 0, plus or minus. It's what most presentday computers use.

But *analog*, on the other hand, implies something quite different: it is clusters of energy resonating in modulations of partly instead of flat *yes* or *no*. It means maybe yes, maybe no, and maybe maybe—it all depends. Certainly the flat yes and no are important. But so are the maybe and sometimes and kind of and it all depends. A sliding analog tone as you teasingly say "Nooooo" to your friend makes all the difference in what you're really signaling. You can put a tailspin on reality by giving your friend a double or triple message.

So there are limits to linearity. Something can make good sense even when it doesn't make logical sense. We find this happening constantly in ordinary life. Jokes for instance are not linear, yet they make a weird sense. Jokes take a situation through several cycles of an iterating pattern with escalating intensity until a paradox skews the cycle and pulls you up short and laughing. You can easily tell a funny joke from an unfunny joke—in other words, you make a quick judgment on its analog *quality*. It's not just jokes, either. Lots of events contain a gut-level patterning and coherence that you just can't explain in the narrow confines of cause-and-effect logic. It's because they are not strictly linear. They are analinear.

But in the everyday world around us, in the material, graspable, measurable landscape of bolts and levers and roads and buildings and eating dinner, analogs operate in such a tenuous, shifty, intangible way that we actually *experience* them rather than think them through. In a strange way, emotion is always conveying the quality of life's analog fit rather than crouching down to measure it with the linear ruler of quantity. Even our senses concur. We do not think—"Ah! 780 nanometer wavelengths of light. Logic tells me that this is red!" Instead we experience it as the shock of red blood flowing, a toreador's cape, a stop light, a lipstick kiss on the cheek.

Our lives are analinear. They reckon quality too. Mere quantity becomes meaningless after awhile—whether it's sums of gold or bedmates or days to live or real estate or bottles of beer. You already know that the thread of your life is not just linear. Consider—it has taken you to places you never expected. It has tangled your personal life into designs beyond rational explanation. The way of the Tao cannot be spoken. It must be lived. Its iterating process states your life again and again in slightly different variations until a theme begins to emerge . . . if you're looking for it. You can get a gut feel for it at a depth that you can never explain.

You can understand analogs even when you can't grasp them exactly by the rigid handle of linear logic. Philip Davis and Reuben Hersh put it this way in *The Mathematical Experience:* "Analog mathematizing is sometimes easy, can be accomplished rapidly, and may make use of none, or very few, of the abstract symbol structures of 'school' mathematics. Results may be expressed not in words but in 'understanding,' 'intuition,' or 'feeling.'"

I expect for instance that you'll quickly get an intuitive feel for the theme of this book even though your reading eye must follow it more slowly through the pages, along the sentences that are lined up on shelves of paragraphing. You will parse the individual sentences long after your overleaping intuition has grasped its central idea. Likewise, you don't have to understand every point to get the point. A sudden Aha!—insight!—gestalt!—a light bulb turns on in the brain—"I see!"—this is what grasps the holistic pattern. It sees how the parts fit, not just the individual units of alphabet lined up on shelves in the warehouse of words.

Analogs discuss the proportional fit and shift of things. In their subtly flowing fashion, they form ratios of comparison, not just final sums. Linear number goes, "Okay, 2 sheep + 2 sheep = 4 sheep." Bottom line. We're counting sheep units while going for the goal, the solution, the concrete answer to the problem. The big 4. These sheep may be of varying ages, sexes, breeds, colors, but all that doesn't matter when we're just counting sheep. All we care about is, "How many sheep?" For the moment, they're just space-filling units of the same sort—and right now it happens to be sheep.

Sheep are the contents being carried within this form of the 4, like lemonade carried in a glass pitcher. But the 4's contents can vary while still keeping its same form—4 sheep, 4 pencils, 4 Empire State Buildings, just as the pitcher can carry lemonade, margueritas, beet juice, beef stew, shit, or 10-penny nails—whatever. We're just doing sums with the units to get a final number. So this 4, when it's used in the linear way, holds merely a quantity of units. A sum. Discrete. End-stopped. The goal. A lump of unitized solution.

But analogs don't emphasize the end goal, the final lump, the *quantity*. Instead they discuss the *quality* of relationships along the way. They bring up all kinds of hinted-at connotations and resonant associations. They open the door to ongoing process rather than closing it down into a summary answer.

And that's the whole trouble with analogs, from a logical point of view. Analogs are relational. They are connotative rather than denotative. Networking rather than end-stopped. They engender resonances that linear logic doesn't want to encourage because it prefers to stay tidy and neat and hurry to a nice solution—not trigger a whole networking throng of related resonances.

Yet analogs do just that—trigger. They provoke sympathetic vibration, whether you want it or not. You can see this happening when the film flickers, blurs, and breaks in a movie theater; you feel it happening when your car develops a shimmy in the front end at 55 mph; you hear it happening when your stereo speaker flutters as you turn up the volume. Or at a football game when enthusiasm drives the crowd to resonate as a single roaring beast all vibrating together. This is the danger and delight of mob psychology: analog connection melds the many into a huge organism vibrating as one. *Analog numbers resonate in networks that reinforce entrainment.*

And this is the whole key to the analog. It does not care about the summary quantity of things but rather, their quality along the way. It makes relative comparisons in a process that is shifting, changing, not striving for an end but rather for the consummate trip, so that we finally never get there, because *there* is irrelevant. To quote that master of analogic, Gertrude Stein, "There isn't any there there." There becomes no goal at all. Instead, you just keep traveling.

Analogs merge with linears to make analinear systems. They shape our body to keep its proportions in workable relationship, so that your feet are big enough to hold your body up but not so big that the platform becomes too cumbersome to move around. Analogs shape a tree so that its branches do not uproot the trunk with an overbalancing mass of leafy green sails that catch the wind.

Before computers, the analinear realm was pretty much a mystery. Its cycling, evolving shifts were—wow!—just impossible for our math. When Mitchell Feigenbaum in the early 1970s started to work beyond the fringe of linear respectability, he was warned that even trying to comprehend such a system was too frightening.

But computers have changed all that. Nowadays they spit out the necessary tedious iterations that would drive a mathematician crazy with boredom and overwork. Finally we can embrace the analinear realm in its huge buzzing messiness of networking connection.

The analinear world flows in continuous waves of events. It is maliciously difficult to break down into sequences and analyze, because it won't stand docilely still. How do you define and break up the rhythm of a three-year-old whirligigging on the lawn? Or a wave curling against the sunset? A bird song at daybreak? You just stand there and watch and listen, and walk away with the memory of something that will never again repeat exactly that way, in just that space and time.

And what did you see or hear? Only rippling motion. Cascading sound. Even with your camera and tape recorder, you can't hold onto the real event. You get a pale facsimile. A memento. So when we stand on the lawn and watch a child play whirling dervish or pause on the beachhead to watch the waves or in the driveway listening to a bird song, we experience cycling process more than end product. We become interested in the dynamic flow itself instead of a static conclusion. Sure, we do pick up a shell or feather or photo to commemorate that time. But we can't walk away with a whirl or a wave or a bird song—only the feather and the memory.

In learning to savor the analog in nature, we go with the flow. We enjoy change. We become surfers riding the endless wave rather than beachcombers carrying away dead matter as a tidy icon.

A relational quality is inherent in the analog domain. This is how our human ties get their amazing power to pull us along by the nose and wallet and heart strings for years, for a lifetime. It is what attracts and repulses us beyond logic and sends us hither and yon across the face of the earth. No, it is not logical that we fall in love or into despair or begin a quest for the Holy Grail, but because it happens, our lives are changed. The analog power grips us. It engenders, it demands, it sustains our human connection with each other.

Some would point out that analog holism is characteristic of the right brain, while linear cause-and-effect logic is left brain. True enough. But these two modes are bigger than our brains. Here on earth, our brains have only recently developed two hemispheres. They are a microcosmic rendition of a larger principle. Fortunately, they give us a means to apprehend these two modes of number which are inherent in the cosmos itself.

Numbers hook up to create the patterns of the universe. They form networks of resonance in the timing and spacing of matter and energy, as well as count discrete sums to quantify the units of whatever is being timed or spaced. Together—as the analinear—they give flowing, connective quality to the universe's discrete quantities.

Your intuition can sometimes pick out a submerged repeating pattern in apparent chaos. In perceiving it, you use the simple skill of pattern recognition—you know, what you did as a child when you studied a drawing and found those six cats hidden in the tree. You found their patterns in space within the time you looked.

Life has infinite patterns hidden in space and time. They cycle in dynamic variations on a theme, nesting the little patterns into bigger patterns into bigger-still patterns that are not merely logical, but nevertheless exist, rather in the quixotic way that the Bluebonnet state of Texas holds the town of Rosebud that contains Bluten's department store where Iris, the manager, displays some wallpaper rolls with a fleur-de-lis design. Wheels within wheels within wheels all cog together to iterate a floral motif in variations big and small.

Analog and linear. Circle and line. Recently I had a wordless dream where discrete linear number was going straight to a solution, but analog number went round and round, reiterating the process. And it's true. Linears sum it up. But analogs play it again and again, resonating over and over. Play it again, Sam.

Analinears take this cycling on to a new level. They get tired of the same old tune. So they jazz it up by flowing into variation, using a line that drives somewhere new. It no longer is just the circle iterating itself, nor the straight line driving to a goal. Instead, it is an old theme that nevertheless takes off in a new direction. Now it's "Play it again, Sam, but with a twist. Jazz it up. Go somewhere new. Take it where I've never been before. And make me thrill to hear it."

Nature plays a noodling jazz that evolves cycling rhythms big and small. It uses analogs to provide the connective beat, more interested in how things hop about in cycles than what they finally add up to. The linears are left to drive the notes home. And when you put the cycle and the line together, you get the analinear spiral.

The spiral is the supreme symbol of evolution. Combining cycle and line, it epitomizes entrainment *plus* change; it replicates an old movement yet drives somewhere new. This spiral archetype is visible in the wheeling galaxy, the cyclone, the dirty water funneling down your sink drain, the DNA in your body. The spiral spins the cosmos together and knits us into being.

And DNA uses not just one spiral but two, bonds them into a double helix to form a supersystem that is doubly entrained, doubly fail-safe in its paired protection of genes. We shall discover that this same plan also appears in the 5,000 year-old I Ching. How can that be? Well, the genetic code describes the solid stuff of organic matter, while the I Ching charts the subtle nuancing mind. West sees tangible quantity; East favors intangible quality. They take up two very different stances on the same bedrock of an amazing co-chaos paradigm. The modern West found it and said, "This DNA builds organic matter," . . . while the ancient Chinese found the same plan and said, "This I Ching shows the evolving mind."

They bespeak very different world views. It set me to thinking that the ancient Chinese must have been very analog-tuned. As I looked, I began to find this confirmed everywhere. For instance, a passage in ancient Chinese is remarkably relational. In fact there's no punctuation, so it's all just one sentence. There are no singular or plural distinctions, no tenses, no articles, even no prepositions or conjunctions in the Western sense. Instead, a web of meaning, highly nuanced and pregnant with connotation, bonds each word to its neighbors in context. The grammar of negation, for instance, is much more relational, keyed to a particular query, and it mirrors the specific question in its response. Thus each word quivers with analog relationship.

Analogy is imbedded into the very brush strokes. Chinese writing originated by symbolizing objects. The first principle of character writing is called "Imitating the Form." For instance, here is *man:* 人. See him standing on the ground with his arms down at his sides.

To portray more abstract notions, a second principle developed called "Pointing at the Thing." Here is the man again, but now with his arms extended to symbolize the character *big:* 大. Now he's showing us the measure of something big, something invisible . . . perhaps he's describing the one that got away.

Thus an image could be modified by a slight twist of the brush to suggest new twists of meaning. Here's the symbol for *mouth:* 口. It is a square opening depicted with the loose analog flow of the brush rather than the sharp linear nib of a pen. And this mouth, when it is changed yet again, becomes the abstract idea of a *word:* 言. Dashes above the mouth show a sketchy version of what once was an extended tongue busily in motion.

This progression demonstrates how the Chinese evolved their written language by modifying what was originally a simple imitation of form. It used pictorial analogies. And how shall we draw something so abstract as trustworthiness or good faith? Why, we'll have that man stand by his word, literally, to honor it. Here is *good faith:* 信. It is made of the man in a radical position standing fast by his word. It illustrates the third principle of Chinese writing, which may be translated as "Joining the Meanings." It links one picture with another to suggest a new condition. Western languages do something similar in what we also call "images"—for example, "He's leap-frogging the issue"—but our Western images are metaphors sketched in by alphabet groups rather than by ink drawings.

Thus Chinese writing began with the imitation of form—using pictorial analogies—although it eventually developed into six basic principles that even included a phonetic harmonizing of sounds and a borrowing of sounds between some characters. All of this evinces a non-Western style of thinking. Indeed, laboratory testing has shown that the Chinese pictograph system activates different parts of the brain from those energized by Western languages based on phonetics. And of course Chinese characters cannot reveal their phonetic pronunciation in the same way that alphabet-based words can. But as a plus, people from the opposite ends of China can read the same language even if they are speaking in dialects so different that they are unintelligible to each other.

Tonality is another characteristic of Chinese. Tonality, by its very nature, is analog: it compares the vowels along a range of tones. Why did this tonality develop? Because it had utility. Since the monosyllables that make up Chinese offer a more limited range of sounds to designate meanings, the language must use varying tones to help differentiate the words.

So many words with the same pronunciation will bring a more frequent punning into the language. The Chinese do not regard puns in the same derisive way that Westerners do—the analog brain dotes on puns that the linear brain groans at. The Chinese language in fact plays sound off against image with a poetic dexterity, and through it, fosters a heightened sensibility toward imagery in general. And since so many words sound so similar, the Chinese are continually "writing" some character's image on their palm with a finger so that the listener will literally "get the picture."

A Chinese man once told me that he wanted to name his child Listening to the Sound of Rain in the Pavilion. That's an incredibly long given name in English, but not in Chinese, what with dropping out all the prepositions and articles and even that unnecessary word *sound*, since it is automatically implied when you're listening to rain. His fascination with this name lay not just in its succinct shorthand, but also in its complex evocation of humanity robed in nature. Most of its meaning is implied, not in the actual characters.

Such subtle imagery is quite alien to the Western style of naming children. But more analog societies tend to call their children, their homes, cuisine, streets by an evocative imagery that is rooted within nature's rhythms. Most Western names, though, have long ignored nature's call and have generally forgotten their old roots in meaning. In fact, in the urban-slick 20th century, names too close to nature became rather quaint and unfashionable—Pearl, Ruby, Iris, Forrest, Garland, Running Deer, Ion, Daisy, Eagle Feather, August.

Chinese culture embraces the analog in myriad other ways, too, not just in the past, but even today. Food, for example, is eaten with a pair of chopsticks that are brought into continuous relationship—using an analog *both-and* style of pincer action. Moreover, people prefer to apply these chopsticks while gathered round a circular table whose central dishes are shared in common, choosing a bit from this platter of poetically named Dragon and Tiger (snake and cat) or that bowl of Phoenix Feet (chicken claws).

Chinese food is precut, so you don't have to bring struggle and disharmony to the table by divisive cutting. Every reach for a tidbit brings one back into continual positioning in the food dance. The chopsticks weave a ballet of flux which acknowledges everyone's place in the warmth of the sharing group. People mostly discuss the food—its appearance, flavor, preparation, purchasing, and so on. People joke about food, remember past meals, moments connected with food. The table flows into the participation mystique. All the culture's children gather around the great analog mother for dinner.

But Westerners, on the other hand, maintain separate plates and identities in a statement of implicit territoriality. They even employ the more binary knife and fork to consume an individual portion of food in a separate bailiwick of plate and place mat. Their knife and fork use a cut-and-thrust of *either-or* division rather than the *both-and* pincers of the chopsticks. Westerners argue and debate far more at the table, too, probably because they are much less bonded to a group identity and to the sacred quality of sharing this communion.

The group is the basic unit of China, just as the individual is it for the West. Analog connectivity in China outweighs the single discrete unit, so that in China the balance among things becomes more important than the things themselves.

Folk events offer major data for a cultural anthropologist. China still uses the lunar calendar to celebrate those major holidays closest to its collective heart—Spring Festival and Mid-Autumn Festival. These three-day fests are agrarian-based, tied to the analog cycle of the moon, unlike most American and European holidays, which are based primarily on individuals (almost always men). But even they are often adapted from old analog, nature-based celebrations such as winter solstice (Christmas) or spring equinox (Easter).

The Chinese lunar calendar is based on the moon's waxing and waning cycle, so suggestive of the feminine that in most cultures, menstruation is named for the moon. This queen of the night. Its silvery mysterious glow receives and mirrors back the sun's light. Thus it manifests yin energy rather than the yang energy so evident in the direct rays honored by the solar calendar.

Yin is the essence of relationship. It has the strength to receive, hold, sustain, background, echo. It is noteworthy that only in the dark night, jeweled with its constellations of stars, do we begin to realize just how much bright Sol blinds us with daylight. Likewise, it is only when we range beyond the bright shield of the ego and slip into the dark hinterland of the soul that we can view the glorious patterns of meaning that wheel beyond the sun of linear logic.

In ancient China, yin and yang were partners, not enemies. Both poles were deemed necessary to each other's existence. *Both-and*, not *either-or*. So in Chinese thought, the mutually exclusive shunting gates of binary number never really took hold. China also never became enamored of the sharply-defined ego that asserts so much personal identity in the West. Even today, most Chinese favor the Marxist-Stalinist line that considers psychology to be a Western pseudoscience, perhaps because they still don't separate themselves from participation in the magical flow of nature. They do not notice the borders of the oceanic unconscious because they still live so much inside it. "Am I dreaming the butterfly, or is it dreaming me?"

By not seeing nature as merely binary, China also avoided the West's split between lofty mind and sinful matter. Joseph Needham puts it this way: "Europeans suffered from the schizophrenia of the soul, oscillating forever unhappily between the heavenly host on the one side and the atoms and the void on the other; while the Chinese, wise before their time, worked out an organic theory of the universe which included nature and man, church and state, and all things past, present, and to come."

Make no mistake, early China was far ahead in technology. It developed many inventions that the West came upon only much later. Gunpowder for weaponry, for instance, developed as early as the Sung Dynasty (420–479 A.D.), long before the West. And by that time, its silk weaving industry had also reached such refinement that its textile machinery far exceeded the West's According to the *Shihuozhi* record of Sung history, "China had invented animal- or water-powered spinning machines with as many as thirty-two spindles capable of producing thirty to fifty times more than manual spinning wheels did . . . the silk submitted by various districts for taxes amounted to 3.41 million bolts."

China invented the compass, gunpowder, paper and printing long before the West. These four key inventions unified and homogenized the culture. Moreover, a shared written language allowed printed edicts to be read and understood across the huge nation. This very early allowed central rulers to impose a strong and unified national identity over the vast land. In a continuous gentle curve from the 6th century B.C. through the 18th century, scientific and technical development rose gradually through the recycling of old traditions in slightly updated ways that allowed a slow evolution.

We might imagine something similar for the West if Imperial Rome had never fallen, but instead, had extended its mighty umbrella over government and culture for the past 2,000 years. China lived in such coherent stability for so long because its feudal system was quite different from Europe's. Instead of highly fragmented and competitive societies pitting their diverse values against each other in weak relationship, China chose unity in government, language, and values. Owen Lattimore says that "ever since the first emperor of the Qin Dynasty conquered six other states and assumed absolute authority [221 B.C.], the form of a unified and centralized feudal universal state had always predominated in the social structure of China." In fact, he suggests that the Great Wall of China was built to keep the Chinese culture contained and coherent in its way of life as much as wall out the nomadic barbarians. Walls both physical and psychological kept away the perturbing dissonance of stray cultures.

For a long time, this Chinese mindset brought a benign balance to the culture, along with steady progress. China became the "Middle Kingdom" in the center of the world, or the only world that mattered. From border to border in this huge land, its people looked down on foreigners as being less subtle, less cultured, less aware and evolved. But in that very superiority lay the seeds of decay.

Meanwhile, on the other side of the globe, after the fall of Rome, there was no longer a common writing or government or road system such as existed in China. So instead of mode-locking into analog bonds at any cost to develop a unified civilization that rose in a continuous gentle curve, Western culture did the opposite. It went binary, turning *either-or*. It split and then split again and again. Continually differentiating, competing, jousting for small fiefdoms of increasingly individualistic clout, this mindset championed linear conquest over analog relationship The West's highly fluctuating pattern of cultural development and invention leapt forward with the Greeks, but then it plunged during the Dark Ages, rose again with the Renaissance. Its many nations were far enough apart to develop many different languages and cultural emphases, but they were still close enough for frequent cross-fertilization.

Since Europe was so far from equilibrium, and so dynamic in that very tension, fluctuations amplified into structure-breaking waves which accelerated into ever-more frequent escapes to higher states of organization on the logos side—but with chaotic feedback erupting from the ever-more repressed analog network. By the turn of the 1600s, the West surged well ahead of China in science. Small fluctuations in the various European cultures created many competitive splits moving toward more and more frequent inventions. The unbalanced system drove easily toward proliferating diversity.

This whole thing makes for an *either-or* mindset in the Western collective psyche. The splitting displays itself in many cultural traits. For instance, it appears in economics as the West's pattern of acceleration and slump, swinging wildly in the boom or bust of runaway change. This has even been called "boom-and-bustiness" in a catchy phrase that somehow holds embedded in its very phrasing the linear view of "booming" male expansion which subsides into an oppositional "bustiness" that implies trouble in female form, holding within its dark attraction the specter of evil times. This boom/ bust flip-flop historically describes the cultures of Austria, Poland, Germany, France, England, Portugal, Holland, Greece, Italy, Spain, and of course, the United States.

This huge momentum toward cascading change in the West is everywhere evident. Alvin Toffler's book *Future Shock* chronicled the accelerating rush of technological change. The perturber has been the enormous tension generated by the deep split in the Western psyche. The yang of linear logic has devalued the cycling analogs of yin. The goal was to win, come out on top, triumph over nature, women, dusky races, the unconscious, sin, whatever subliminally smacked of that dark analog domain.

Yang won. Consciously, that is. But yin sank into the repressed layer of the culture. Its energy became not just dark, but dangerous, sinful, forbidden, misshapen. The act of repressing yin has fostered such momentous swings in Western history as the rise of logic and male patriarchy in Greece; the Dark Ages where a huge priestly population forswore the sinful daughters of Eve; the Crusaders who sought a visionary jeweled chalice called the Grail—such a feminine symbol—even as women were locked away into chastity belts; the solitary alchemists who tried to transmute lead through the inspiration of a *soror mystica* or mystic sister; the Renaissance which momentarily brought men and women back into cultural equilibrium and by that, reaped a new birth; the wholesale conquest and rape of that huge green virgin called the New World; the Puritan age of Bible-thumpers and witch-burners; the linear white society's smug suppression of more analog, dangerously darker Amerindians and black slaves; the Industrial and High-Tech revolutions whose supremely logical bounty has also poisoned Mother Earth.

This past is airbrushed into the subtext of our culture. Look at the main hero of 20th century entertainment: a solitary macho hero who has lost or damaged or betrayed or shot or forgotten the girl. He's the dick who climaxes to a solution while getting over some dame—really, how Freudian can you get?

Entertainment tries to escape our linear shackles. Our pop music rocks and rolls, our jazz noodles instead of hewing to a straight line. The White Lady of heroin, snow, smack whispers to get it on. Get down. Get wiped, stoned, wasted. Get bad. Get crack. All these fissures into the dark reveal the culture's deep split.

Western culture fears that huge dark power hiding deep in the cracks. It backs off from yin, shouting, "No! You're too witchy, irrational, entrancing. Too underground, underworld, undercover. Your lure is dangerous. I *must* not surrender." So of course it succumbs. Into drugs, drink, crime, violence, dazed cults.

The West has conquered nature, and by it, has turned her into a raddled whore. The split finally pushed us to such extreme alienation that it left Mother Nature raped and crippled by technology. The tormented culture "subdued" nature, viewing the analog dark feminine as too overwhelming and unbridled, demanding to be controlled by the masculine. All this culminated in a technological threat to the human species, to other species, to the planet itself.

And finally, we are realizing that going so linear is dangerous business. An *either-or* mentality that champions bright yang cannot get rid of that dark other pole, but only suppress it. To have one's way with nature, to possess it, control it, own it—that is Western. A foolish goal, finally impossible. In having it, we get had. But to hold nature intact, within and without—that is Eastern.

As the West busily boomed and busted, meanwhile, inside its Great Wall, China went on with its centrally sanctioned inventions and very slow, steady rate of development. Feudal society became so ultra-stable that it resisted change. China became Confucian, honoring a stability of form over the Taoist tolerance for chaos that erupts into creativity. China became so set in recycling its analog habits that it forgot how to incorporate linear change. It could no longer change without threatening its very identity, the nation's way of conceiving of itself.

So Chinese technology lost out to the driving linear force of the West. In the last 350 years the West jumped well ahead in science and technology. Why? It began to apply abstract logical theory and scientific experiment to all sorts of areas, dropped its feudalism, and connected its capitalism to technical development. Jin, Fan, Fan, and Liu (all graduates from universities in China) ruefully chart the Chinese perspective on 2,500 years of 2,000 technical achievements in Chinese and European history. Their essay called "The Evolution of Chinese Science and Technology" sees Chinese science as showing plenty of early promise but failing to live up to Western technical progress in the long run. Instead, China stayed awash in the cyclic tides, turned symbolism into superstition, became imprisoned in a great wall of inertia. In brief, China went too analog. Too steeped in cycling tradition.

It was also an analog love of tradition that codified the masculine and feminine roles into collective sexist ideals of war lord and foot-bound lady. The government itself eventually took on a persona of war lord, yang and dominant. It maintained iron-handed rulership over the pliant yin masses. Severe laws could not be challenged without unsparing punishment. The huge submissive populace stayed feudally isolated and dependent on its rigid yang government for the sole definition of the national identity, like some enormous foot-bound lady who is kept hobbled and submissively at home by the lord and master of the family. John Fairbank says in *The Great Chinese Revolution: 1800 to 1985*: "Authority figures should not and could not be challenged. Criticism endangered authority and was therefore unacceptable."

Ritualized yang-yin roles crippled the culture's ability to evolve as it forgot to honor that essential seed of the opposite within. In a modernday metaphor, a man's psyche possesses its feminine component, just as a woman's holds its masculine aspect. For wholeness, one needs to see, honor, and develop that contrasexual complement within. We all carry—and need—both yin and yang energy.

China feared change because it seemed synonymous with chaos. This profound, ancient culture became so equilibrated that for nearly two thousand years nothing could knock it out of its ultra-stable pattern and move it onto a faster track of evolution.

But the 20th century finally broke that stasis. The 1949 Red coup, the Cultural Revolution, Tiananmen Square 1989—through all this, China has struggled hard to modernize. Yet change still feels cataclysmic, alien and fraught with danger. After the student uprising in Tiananmen Square in 1989, families were told to betray to the government any members who had deviated from the official line. It echoes an earlier imperial China: "For the sake of the Great Cause, destroy your loved ones." Students wrote compulsory essays on the danger of chaos. Faculty and students at Beijing Normal University attended political study sessions to "turn around people's thinking." *Newsweek* reported that a dean spoke the verbatim words of countless official editorials when he declared, "We want to make sure everyone's thinking is unified."

Despite China's abrupt leap into linear materialism wearing the red shoes of Marxist ideology, the culture still is deeply analog. The common folk still do not view their nation in a boom-and-busty way, but rather in the ancient cyclic sixty-year tides that have rolled through more than 5,000 years of recorded history. China still restricts travel beyond the womb-like boundary of the motherland. Its government takes such pride in being "revolutionary" that no one can revolt against it. The worst tag that the ruling clique can hang on a dissident is "counter-revolutionary." Thus China has ritualized even the shock of revolution into a safe tradition, has embraced "revolution" in the name of the approved regime and so has robbed the word of any meaning, swallowed it whole in a yin-like way. Shades of double-speak, of the *both-and* mindset clutching all in its embrace, of the smother-mother.

Nevertheless, modern China becomes continually more aware of economic benefits in promoting personal identity, cultural exchange, and freedom in general. I admire much about the Chinese people and hope for their emerging opportunity to embrace the best of their past in their future. Essential to this development will be overcoming their fear of losing containment. They have already debunked the dread that saw "foreign devils" as ghosts with white skin, eerie blue eyes, and diseased yellow hair. The East is beginning to reach out more and more toward the West for its technical skills, its higher standard of living, its mod-pop zest. But it must also learn to discern the difference between Western talents and its bad habits.

The West reaches eastward, too. Techno-weary Westerners grow starry-eyed over the mysticism of ancient China. Attunement with cycling rhythms. Embracing all reality, accepting all, digesting all, surviving all. Sounds grand, doesn't it? But eventually so much analog containment became too exclusively yin in its embrace. Passive. Accepting. Absorbing the increasingly ritualized procedures without question or protest, overwhelmed by the weight of accrued cycles. Shouldering the burden of the past, the history, the ancestors, the government. A Chinese man once bitterly told me, "The trouble with us Chinese is that we're too patient. We'll accept anything."

Thus going analog does not necessarily mean a kinder, gentler society. For China it also meant that the society stayed more in the unconscious, iterating and reinforcing its collective rituals, forming a rigid social structure strong enough to contain and stabilize the unconscious masses, including those many unfiltered, undigested psyches going walkabout in the culture.

Ghosts and monsters and demons and spirits abound in analog cultures. Gods and omens and good luck totems. Charms and lucky numbers and auspicious wordplays. Favors and bribes and grudges and gangs and feuds. An analog society lives mostly in the participation mystique, swept up in the overwhelming tides of favor and disfavor, buzzing in the hive mind as expendable drones or worker-bees supporting an Emperor's distant war or Mao's little red book of homilies. Whatever the boss-ego envisions, the rest must identify with. The leader projects some personal inner goal without knowing where its boundaries stop, constellating in the masses its blowsy shifting image unredeemed by enough distance and definition to permit analysis, much less criticism.

Living nestled in the participation mystique also has its strong rewards, though. It can bring harmony with nature's cycles and a magical connection with the eternal verities. The West needs more analog connection, beyond any doubt. It must quit raping nature brutally, must learn to hold nature in embrace rather than have it in plunder. And unfortunately, the more Western that China goes, the more it despoils its natural beauty and tried-and-true virtues. The age-old biodegradable agrarian assumptions don't hold any longer for styrofoam and urban pollution. Habits of thousands of years fall against the brutal blows of technology.

But the unmitigated analog is often brutal too. Not so much to its land, but to its people. The stranglehold of mindless ritual tends to lock people into unconscious roles that sacrifice the individual. It recycles the past without allowing much progress of forward-moving linearity. For example, at the turn of the 20th century, the Ashanti in Africa still sacrificed a hundred men and women to accompany their god-king when he died, so that he would be happy in the beyond and bless the tribe. His harem was especially singled out for this honor of accompanying him into death. Willing victims gathered at a holy banquet and got drunk. Then female executioners came in and strangled the many widows with leather straps.

Human sacrifice. The Egyptians did it on their way to geometry, the Chinese on their way to Xian, the Jews on their way to Moses on the mountain. It's an old, old pattern, this living sacrifice to the gods of analog cycle. Honoring just the analog is certainly as risky as honoring only the linear.

Our globe is much like a huge split brain, long oblivious to its other side, but now finally exploring new connective paths into that puzzling other hemisphere. There is much to hope for in our new global society. We can take heart in considering how the human brain's two hemispheres can cooperate to walk a central path that melds logic and holism. Wedding bright linearity with nuanced analog shadow will give us a fuller, in-depth picture of reality. It has happened a few times before in history—in ancient China, early Egypt, classical Greece, the European Renaissance—and each time, it was a balance of creative flowering poised between yin and yang, joining analog wisdom with linear smarts.

To integrate the analog with the linear is the aim of this book. It is written to engage both sides of your brain. It has both scientific and spiritual topics. If you prefer, just browse what is less-familiar in format, allowing yourself to flow easily on. It will trigger change.

Can we reunite our own brains, our cultures, our globe, have *and* hold earth in unity? Embracing the analinear paradigm is a first step.

WORKSHOP 2

Plan a program that combines discussion and experiences. Select book paragraphs to introduce the following topics—or your own.

Opening ritual.

❣ Bring to this workshop a book with the meanings of names. As people come in, have them look up their names and write the meaning on a name tag. Ask them to introduce themselves by it.

❣ Is your watch analog or linear? Could its presentation of time subtly shape your daily view of reality? Explain your opinion.

❣ Ask the group members to move to music in a line. In a circle. In a swirling spiral. How does each pattern feel? Discuss how your body responds, and what that movement suggests to the psyche.

❣ Imagine that the globe is a huge human brain with the East and West as right and left hemispheres. How might the two sides see and utilize each other's special gifts to work more cooperatively? Is this realistic? From your own experience, have you seen any evidence of this happening? Explain.

❣ Consider women's roles compared to men's. Do you see yang and yin pressures in society? Can it, does it, should it change? Where do you see the yang principle operating in your own life? The yin principle? Which is currently more prevalent for you personally? Can you do something to balance them better in your daily routine? Would it be worthwhile to bother? Explain.

❣ Meditation: One at a time, think of three people you know well and call them up to memory. Characterize each as more linear or more holistic. Next gauge them relative to each other along an analog dial range. Have any moved into a different spot along the dial through time? Discuss in small groups or the whole group.

❣ Invite people to recount briefly the discovery of a larger pattern of behavior in a family or work environment that had been formerly lost in the weave of everyday events. Any evolution?

❣ Discuss learning to see patterns in your own life. How can you practice truing up your intuition? Can you develop ways to gauge it? How can the way of the Tao become more instinctive?

Closing ritual. Five minutes of feedback. Announcements.

The East-West Fork

How did the East and West grow so different upon this common globe? Old China's culture was based on the yin-yang philosophy, which spread throughout the East. Western culture is deeply rooted in Greco-Roman-Judean thought. Both sides of the globe used the concept of duality—it's found in both Plato and the I Ching—but the Eastern version is far more elaborated and nuanced.

To Plato, the cosmos was in a duel—the ideal versus the real. The struggle between lofty thought and hard matter generated everything. Ongoing conflict between Idea and Form churned out the cosmos. Abstract versus concrete. Mind versus matter. In this view, reality was the prize won by polarized enemies fighting to a goal. It promoted an *either-or* mindset. You win, I lose . . . or the reverse. Win/lose. Mind/matter. Black/white. Good/evil. Yes/no.

A century before Plato, though, Heraclitus spoke of this duality, but differently. He emphasized that transcending a single-minded focus on either side of duality will allow you to see the synergistic whole. His *both-and* approach sounds analog, even Taoistic. But the West was not so enamored of this holistic vision, and so explored it far less than Plato's.

Western thought favored the linear divide-and-conquer style of concept formation. Logic-chopping splits things into categories: "It's either this or that." Aristotle especially valued matter and its scientific measurement—divide and conquer that substance. He established the pragmatism of taking the measure of things to help establish an empirical proof. His own disciple Alexander the Great actually went out to divide and conquer the known world. By the time of Julius Caesar, "Divide and conquer" became the motto for success. It was only logical.

But the rich, dark holistic side of life just wouldn't go away, not even when logic repressed it into the disreputable unconscious. So despite a Western swearing of allegiance to logical linearity, it wasn't able to "disappear" the analog pole completely. And sometimes the West still toyed with the mystic thrill of melding duality into a transcendent third condition. The allure of the *tertium quid* or "third thing" slowly grew, perhaps even out of Aristotle's syllogism. There at least a logical product came from the relationship between two ideas, if no transcendence.

Around the 13th century, Christianity settled on its own mystical "third thing." The early Gnostic idea that the Holy Spirit was feminine became a heresy, and the male duo of heavenly Father and earth-born Son were joined by a third male element, the Holy Ghost. It became seen as an intangible form of god's male potency and was often portrayed as a strong wind or a light beam. This line of force beamed down on Mary and impregnated her with Jesus, whose birth bridged that painful gap between the immensity of god and the frailty of our human condition. Jesus became a psychological stairway to heaven. In this doctrine of the all-male Trinity, we can discern the model of the transcendental third thing.

Philosophy also tried to resolve the tension of *either-or* shunting that plagues Western thought. Kant spoke of the antinomies, which Hegel developed into the principle of thesis—antithesis—synthesis. Yet that too wasn't nearly so thorough-going a model of polarity synthesis as we can find in the I Ching.

From the beginning, Chinese philosophy saw the cosmos as organized into unity, duality, and a transcendent third principle of change. This synergy is symbolized in the Tai Chi.

Although this Tai Chi symbol is familiar to many, some may not realize that the white stands for an emerging foreground, while the black stands for its containing background. Basically the Tai Chi depicts the modern idea of active field and passive ground. Yang and yin are complementary necessities to each other. Bright yang cannot actively emerge without having a dark background; yin cannot passively contain without having something to hold. One cannot exist without the other. They do not fight as in Western thought; instead they enable each other to exist.

The Tai Chi's dot of white within black, and of black within white symbolizes the dynamic potential for the forces to change into each other—meaning that the background can rise into foreground by giving it attention, as meanwhile the foreground will recede into the background of attention.

Consider the familiar vase/faces of gestalt psychology. You may suppose that in this image, white symbolizes the actively emerging yang foreground, while black symbolizes a passively holding yin background. But that isn't quite the case, for if it were so, then you would only see the white faces and never the black vase.

Yang-Yin as Field and Ground

This black-white color coding of yin and yang is actually just a convention that dates from the old days of China before the depiction of motion was possible with film, television, computers, and such. In those static old days, white conveniently symbolized an emergence into foreground, while black symbolized a retreat into the background. But the truth is, if your eyes are focusing on faces, then the faces are yang; but if your eyes are focusing on the vase, then *that*—the vase—is yang.

Hermann Hakan notes this alternation in patterned chaos when he says that there are two attractor points—vase and faces—and the

viewing eye switches back and forth between them at an essentially unpredictable rate. Vase and faces are merely the attractor points that sit in the two wells of possibility.

Bifurcation into Two Wells of Possibility

If you alternate your focus between them, you shift between yang's emerging focus and yin's retiring background:

| yang | yang | yang | yang | yang | yang |
| faces | vase | faces | vase | faces | vase |

Yang-Yin Shift in a Dynamic Gestalt of Field & Ground

This shifting focus is what is implied by the dots of opposite color in the Tai Chi—it represents an ongoing flow of perpetual energy that is polarized into alternating between field and ground.

If we see this shift of attention within the black and white Tai Chi symbol itself, and if we view it as a series of successive frames, the ball even seems to be rolling and turning.

Yang-Yin Alternation Seen as Motion

In fact Leon Glass and Michael Mackey in their book *From Clocks to Chaos* present a chaos theory diagram of the solutions of a differential equation in a Period 2 limit cycle oscillation that looks uncannily like the Tai Chi. Its vector forces turn into the Tai Chi symbol.

Vector forces

Yang is really whatever stands out to grab your attention. Yin is whatever contains it. And when your mind eventually realizes that both partners together create the whole—as in the Tai Chi—then you achieve transcendence over the parts.

Duality in the ancient Chinese system means two complementary conditions which transform into a higher level of unity by transcending the old polarity. First comes a state of undifferentiated unity. Things seem all alike, just a gray blur ●. Slowly you begin to differentiate between this and that, and the gray blur starts to define itself and move from ● into ◓.

Ah, comes the dawn and we see a horizon separating heaven from earth, for nature analogy abounds in Chinese thought. Bright heaven and dark earth become equal and complementary powers. As this happens, we realize that these two complementary states define each other: black holds the white; white emerges from the black. But they also keep shifting places as foreground and background and thus they create the Tai Chi symbol: ☯. Both poles are necessary for movement, since development comes only through change.

You can see this mode of thought in the ancient Chinese story of the man whose horse ran away one day. The others in the village lamented and said, "Oh, your horse has run off! Too bad." But the man just shrugged, "Maybe yes, maybe no." The next day his horse came wandering back home with two wild horses tagging along behind. The villagers rejoiced, saying, "Look, now you have three horses! How lucky you are!" But the man shrugged and said, "Maybe yes, maybe no." The next day his oldest son, while trying to tame one of the wild horses, fell off and broke his leg. The villagers said, "Oh no! Now your oldest son has a broken leg. He can't help you in the field. How unlucky for you!" But the man shrugged and said, "Hmm, maybe yes, maybe no." The next day the emperor's army came through the village, conscripting all able-bodied young men, and of course they couldn't be bothered to cart along the son with the broken leg . . . .

How far shall we carry this shaggy tale forward? It is still told. A 1994 television episode of *Northern Exposure* offered it as a story told by Marilyn, the Native American medical receptionist. This old tale could go on forever. It actually does in life. How many times has a piece of seemingly terrible luck turned out of great benefit for you, or at least better than expected? The I Ching philosophy suggests that no issue is merely binary. Sure, things may seem black or white at first. "Either it is or it isn't. That's all there is to it!" It seems . . . until the situation evolves.

Nothing remains so simple, including the I Ching. So remarked Kang Hsi, an emperor renowned for his wisdom and generosity who ruled in China from 1662 to 1722 A.D. "I have never tired of the Book of Changes, and have used it for fortune-telling, and as a source of moral principles. The only thing you must not do, I told my court lecturers, is not to make this book appear simple, for there are meanings here that lie beyond words."

Seeking to hold both poles within your attention will allow you to identify with more than half of an issue. You transcend the two attractor points that pull at you willy-nilly, each demanding loyalty for only its side. In psychological terms, if you can hold both sides of an issue in your mind, seeing ill and benefit rather than labeling one pole as automatically good and the other as therefore evil, then somehow this balance allows you to move calmly past a linear label of good and evil, demons and exorcists, saints and sinners, white hats and black hats, villains and heroes.

Then we become neither perfect nor perfectible, so we can seek wholeness, knowing that it means acknowledging the dark side, learning to befriend rather than be-fiend the shadow. It will transform and support us instead of draining us in vampire fashion. Then we can perceive more clearly the flux of energy surrounding each attractor point in an issue; we can transcend its polar pull to evolve a new attitude that embraces both dark and light as beneficial. And this changes the reality itself. It is walking in Tao.

Transcending levels is the dynamic of the Hindu-Buddhist chakra system. The chakras are associated with glands located at seven sites along the spine. Two lines of complementary force run up the spine through these seven transformer stations to reach a transcendental union with god at the crown of the head. This movement is similar to that symbolized in the old Greek healing rod, the caduceus. Two serpents twine up the staff to grow wings at the very top, depicting final transcendence. These winged serpents symbolize primal power that has risen from the earthy depths to the lofty heights.

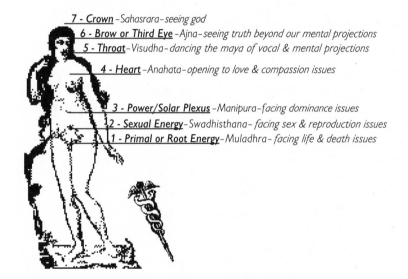

7 - Crown -*Sahasrara-seeing god*
6 - Brow or Third Eye -*Ajna-seeing truth beyond our mental projections*
5 - Throat-*Visudha-dancing the maya of vocal & mental projections*
4 - Heart -*Anahata-opening to love & compassion issues*
3 - Power/Solar Plexus -*Manipura-facing dominance issues*
2 - Sexual Energy-*Swadhisthana- facing sex & reproduction issues*
1 - Primal or Root Energy-*Muladhra- facing life & death issues*

The Seven Chakras

You can get stuck in one or two transformer stations along the way and not energize the other chakras enough. But hopefully, you will move on through a successive development of all the chakras. And if you do, you reach unified behavior.

Another version of this chakra system can be seen in the West as Maslow's hierarchy of needs, where the modernday psyche also rises through the successive levels of development—although in Maslow's version, the psyche transcends not to god but just to a better self-integration. Erikson and others have similar hierarchies.

In Hinduism, the dot of red paint on the forehead provides a quick symbol for transcendence. This "lotus" eye transcends the duality of those lower fleshly eyes set in two different eye sockets so that they can register ● from the angles of ● and ○, and thus give two differing streams of information to the brain. This double and offset vision allows a separate set of data intake for each eye, a duality which the brain processes and then transcends with its unifying synthesis to create the larger vision of in-depth perspective.

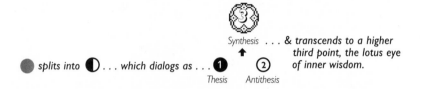

This lotus eye surmounts mere data with insight. It is a powerful concept, this movement from gray confusion to insight through transcending two complementary poles. Let's call them yang-yin . . . or maybe it's yin-yang? What's your preference? Which pole comes first? Neither, actually. They emerge equally and oppositely at the same time. I've noticed that, appropriately enough, the West tends to say "yang-yin," while the East terms it "yin-yang." Here is just another instance of their differing emphasis. The East puts first the matrix of introverted yin, but the West puts first the emerging focus of extraverted yang.

Notice that duality can never be integrated on its own level. It must always rise to an overview with more perspective. There's even an ancient Latin phrase for it—*tertium non datur*—which means that the third position cannot occur on the same level as the polarized data initiating it. Instead it must rise or sublime to a higher, more complex level of organization. This recalls Ilya Prigogine's concept in chemistry of "escape to a higher order," which occurs in dissipative structures. Prigogine's book *From Being to Becoming* succinctly highlights the shift in modern science as it moves away from the linear goal of *being* to the analog process of *becoming*.

The transcendent third position enlarges one's vision until new data accrue from the continual changes in life, and eventually it challenges the old insight enough to turn understanding gray and confused again. This becomes the springboard for a new, larger-scale movement toward clarity. In the picture below, gray obscurity is the starting point for action that seeks mirroring relationship by becoming one pole in dialog with another pole. They integrate again at a third higher level of order. Of course now the scope is bigger. Then the whole process repeats, but this time it's on a higher scale of organization. Evolution continually creates a new transcendent third state, which then grows stale and so becomes the basis for a whole new upward thrust.

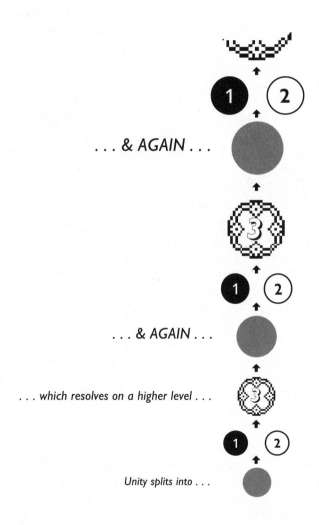

... & AGAIN ...

... & AGAIN ...

... *which resolves on a higher level* ...

Unity splits into ...

You can see this very same "escape to a higher order" portrayed on the statues of Tibet's evolving god Chenrezi, where each loftier head portrays a wider wisdom rising from a fresh insight. Each new face that is spiraling upward represents the next incarnation of the Dalai Lama.

So this third eye of insight is higher order . . . yet it's also restless. It continually seeks completion through further evolution. The archetype of the 3 heralds the birth of a new level that allows renewed growth. Cultures have long noted this archetypal force of the 3. Christianity offers birth into a new life through belief in the Trinity. The ancient Egyptian Copt Maria Prophetissa said, "One becomes two, two becomes three, and out of the third comes the one as the fourth." Laotse described it in the *Tao Te Ching:* "The Tao begot One. One begot two. Two begot three. And three begot the ten thousand things." In folk wisdom, the third time's the charm. In fairy tales, it takes three trials for the protagonist to transform the trouble. And as we shall see in chaos theory, Li and Yorke express this same concept of the transformative 3 in their provocatively titled essay, "Period Three Implies Chaos."

In science, the analog domain is finally stepping forward to take its place beside the linear. But the ancient authors of the I Ching saw it long ago. They were analinear scientists of a sort, working with cosmic number models to develop meshing linear and analog patterns. They, like modern chaos theorists, were interested in the rhythms of spacing and timing that recur in matter and energy with slight variations on scales large and small.

Kristofer Schipper and Wang Hsiu-huei report in "Progressive and Regressive Time Cycles in Taoist Ritual" that "In Chinese thought, the universe is apprehended as an infinity of nesting time cycles that, because of their formal correspondences, may be manipulated as though they were interchangeable . . . ." Thus ancient Chinese study had its parallel with what is happening these days in the science of patterned chaos.

Eastern culture fostered a cosmic metaphysics that ventured beyond the limits of linear explanation. This mindset tends to imagine reality not as a sum of linear units, but rather as an exponential progression. A simple instance of this is the poster celebrating the Spring Festival that I bought from a Chinese street vendor in a dusty rural market. The picture shows three Confucian gods of benign plenty, a corpulent Buddha, and even a golden bat thrown in for Taoist good measure. The style had a wholesome perfection that is appealing to a peasant audience, or so an urban Chinese intellectual rather scornfully informed me. Down one edge of this rosy, jolly scene are written the Chinese characters for *Lucky star whose rays to Earth bring the warm welcome of 100 happinesses.* Then down the other side is written *The bounteous spring engenders 1000 good things.*

Of course, 100 and 1000 are stages in exponential progression, formed not by adding chunks but by raising 10 in exponential leaps, as $10^2 = 100$ and $10^3 = 1000$. This poster motif subtly bespeaks the analog cycling mentality of the Chinese mind rather than a Western mindset which tends to think in discrete, additive chunks. And this poster portrays no far-out intellectual's philosophy, but rather, it reveals the Chinese cultural gestalt at the most basic level. In this embracing analog culture, all three major religions—Confucianism, Buddhism, Taoism—can coexist in harmony on one peasant poster. Where Western religions tend toward divisiveness, Eastern religions tend toward inclusiveness. The West conquers, the East absorbs.

Until lately, that is. The West has begun to note and appreciate analog process. Exponential growth is the very trait that is fundamental to the chaos algorithm and to fractals, as we shall see. And nature does tend toward exponential growth and decay. Look at birth rates, oaks from acorns, yeast rising in bread dough, and so on. Analogs give the cycling quality to nature itself.

The holistic mindset is trained to perceive and describe abstract patterns in the flow of passing data. When I went to China as a university teacher, I found my students peculiarly adept at intuiting symbolic resonances. They were quite gifted at recognizing and sorting out patterns of abstract thought. Yet to the more goal-focused, matter-minded Westerner, this kind of conceptualizing may seem unaccountably elusive and ineffectual, especially during the past few centuries when the mechanics of nitty-gritty Newtonian reality held sway. Those glorious flights of intuitive imagery that formerly soared in ancient China have fallen broken-backed into the gray urban sprawl of its present-day cities. A culture that could envision the cosmic balance of the Tao and manifest it in such striking ancient art and harmonious architecture, has somehow lapsed into a bleak uncultured modernity of 20th-century cities that are notorious for their lack of physical flair.

But China still holds supreme distinction wherever it has been allowed to keep its ancient culture intact and evolving—notably in its glorious cooking and incomparable handicrafts. And as the new 20th-century Western sciences develop their own symbolic thought and wheeling patterns above and below ordinary consciousness, an inordinate proportion of those who make significant discoveries show names of Oriental origin.

Eventually I developed the opinion that dynastic China slowly sacrificed its tracking of nitty-gritty specifics so vital to Western science for a vast lexicon of nature imagery. Symbolism surrounded the emperor, bureaucracy, manners, morals, and all this gradually reified into empty ritual that left the rulers and the masses culture-bound. Revolutionary Red China of 1949 saw the problem, yet its sudden determined flip-flop from stagnant ritual into raw Marxism didn't really solve it. That abrupt swing into dialectical materialism totally repudiated the feudal past, but ironically it also deprived the culture of its richest gift, the deep connection with pattern beyond the material limits.

It's no wonder that China went communist, for group sharing and group identity are completely in keeping with its analog values. But once again China took its gift for abstraction too far, this time believing too wholly in an idealized, rosy-red-dawning of material plenty to be shared equally among all in the beloved group. Still relational, still connective, this new vision dreamed of a sturdy peasant solidarity that was grounded in communal plenty. The society repudiated the intellectual scholarship it had so long fostered. That swing from abstract idealism to material idealism is an understandable over-correction, but bereft of its deep past, China marches stolidly into the future on borrowed Red shoes rather than crafting its own pair using those ancient skills retrained into a modern rendering of its own glorious past. This book intends to honor China's great past and re-interpret some aspects of its analog wisdom in the light of modern logic to suggest an analinear synthesis.

To understand the I Ching, we must delve far back into the Chinese past. This document is obscure even in Chinese. After all, ancient Chinese is not a very exact language. Read twenty different English translations of an old poem and you may well suppose that they are twenty different poems. Each translator personally relives—and must—the moment of creative insight that occurred in the poet.

It is why the best Chinese poetry is short and intensely subjective. And why the I Ching has been translated so many ways.

Thus it becomes vital to read below the details for the underlying pattern that is being described. The imagery holds more than just poetry; it yields analog accuracy. This "truth in imagery" approach bespeaks a very different mindset from the West's. Does a mindset create its own language or vice versa? Most likely, both. At any rate the I Ching sprang from a culture attuned to nature and its recurring relationships.

Let's cycle back five thousand years and then start forward. Let's begin in prehistory before the I Ching symbols were first brushed—or I should say, scratched into dirt with a stick, and then incised more permanently onto the bones of animals—for it is with these scratches onto bones that the recorded history of the I Ching begins.

The I Ching originated in 3322 B.C., so goes the apocryphal story. The Emperor Fu Hsi saw a "dragon horse" rising out of the Yellow River. From this strange beast's markings, he got the inspiration to design the Ho Tu or Yellow River Map. (*Ho* means River, while *Tu* means Map or Plan.) These 55 spots became the blueprint for the I Ching. All this is according to very ancient classical Chinese history, which is hard to distinguish from myth.

The ancient Chinese preferred their rulers to be wise rather than brawny, and Fu Hsi was the first in a line of emperor-sages. He got credit for outstanding intellectual leaps in China, much as Paul Bunyan got credit for superhuman physical feats in the Minnesota North Woods. Thus Fu Hsi is honored with teaching the people how to write, to cook, fish, trap, chart heavenly and earthly cycles . . . rather like a scholarly and public-spirited Paul Bunyan of the intellect. He became regarded as the father of Chinese culture, and his wife Nu Kua the mother. An ancient stele shows them holding a square and compasses in order to take the universal measure of all things.

Later emperors were credited with other cultural developments. For instance, the emperor Shen Nong (his name literally means deity of agriculture) started markets for exchange and barter. He introduced farm tools, agriculture, and animal husbandry. He discovered medicinal herbs and their use in the treatment of illness, and so he is considered the father of Chinese medicine.

All this development went on until the next big leap came for the I Ching, around 2200 B.C. It happened while the beneficent Emperor Yu was cleaning up after the great flood. He saw some intriguing designs on the shell of a tortoise climbing out of the Lo River. Its abstract markings inspired him to create the second part of the I Ching. It was called the Lo Shu (or Lo River Writing), and it was a schema for mobilizing those earlier I Ching figures. It showed how to join the 8 trigrams into 64 hexagrams. A myth? Sure. But the result, as we shall see, exhibits the analinear math of co-chaos.

James Legge describes the Lo Shu thus: "This writing was a scheme of the same character as the Ho map, but on the back of a tortoise, which emerged from the river Lo, and showed it to the Great Yu, when he was engaged in his celebrated work of draining off the waters of the flood, as related in the *Shu*."

All this sounds quite fantastical—and it probably is. Most likely, scholars in the courts really developed the I Ching. They were much like grad students nowadays—they do the work and their professors get the credit. Nevertheless, so many small details of legendary Chinese history have now begun to be verified by recent archeological discoveries that historians are starting to suspect that in general, the dates and locations regarding at least some of this legendary history may be rather accurate, and Chinese mythology often holds a kernel of truth intact.

Symbols scratched onto animal bones are our first permanent record of the I Ching. It seems to have been used quite frequently during the Shang dynasty. We have bones, for example, from about 1250 B.C. with marks on them which are believed to represent hexagrams. During the Chou dynasty (1122 B.C. to 256 B.C.), the I Ching was written down in manuscript form. It was copied and even revised. In his *Records of the Historian,* Szuma Chien says that Confucius read the I Ching written on bamboo slats bound together with leather thongs, so often in fact that the thongs broke three times. In the *Confucian Analects,* Confucius is quoted as saying, "If some years were added to my life, I would give fifty to the study of the I, and then I might come to be without great faults." Modern scholarship, though, considers it unlikely that Confucius ever saw the I Ching, and regards these stories as apocryphal, for the Chinese liked to consolidate their collective wisdom under the aegis of a few respected names.

The oracle itself was at first just a short text that contained the basic hexagram titles, judgments, and lines. It was called the *I* (pronounced E or Ye), usually translated as *change* or *mutation*. But according to Fung Yu-lan in *A History of Chinese Philosophy*, this title really had two meanings—*change* and also *easy*—because this oracle was considered a much easier technique for predicting change than the old turtle shell oracle. Later the name expanded into I Ching. The added word *Ching* means book or classic compilation. It was needed when the burgeoning Confucian commentaries of the Ten Wings were bundled into the short text to swell its volume into a book. Finally, the name *Chou* got prefaced onto the title by historians of the Tang Dynasty to signify that the system was first written down in the Chou dynasty. The Chinese nowadays just call it the Chou I.

As a document of the Chou people, the Chou I describes those times in images and analogies, but it also mentions the place names and rulers of the preceding Shang dynasty, the last of whom was a tyrant notorious for his degenerate cruelty and extravagance. The Shang dynasty itself, however, soon passed away in history, mostly invisible except for its various mentions in the I Ching.

But in the province of Henan, during the 19th century, peasants often plowed up oddly marked bones while tilling their fields. They called these fragments "dragon bones" and sold them to druggists. To the Chinese, the image of *dragon* connotes creative energy and heavenly power. But the folk didn't know exactly how this power echoing from the past originated; they didn't realize that these fragments of scratched bone were actually the old records of oracles from the Shang dynasty more than three thousand years before.

Finally around the turn of the 20th century, archeologists realized that this was actually the site of the ancient Shang capital. They started to dig and found Shang tombs and tools and even artful bronzes. By 1910, scholars deciphered enough of these old scratched bones to find they recorded the oracular divinations of an empire so ancient that it had been forgotten everywhere except in legend and in the I Ching, where the brutality that marred the decline of the Shang dynasty is bewailed in some verses. These deciphered bones corroborated the ancient saga of the Shang and Chou dynasties that was carried within the I Ching.

Let's consider the philosophical mindset of ancient China as we can see it in the I Ching. Consider what it meant to live in a world circumscribed by a foursquare earth and a limitless encircling sky, set in a domain holding one smack in the middle of things, no matter where one stood. Snug in a landscape that eventually became called the Middle Kingdom because it so obviously was the center of everything. Everything that mattered.

Why a foursquare earth? Because we stand on it and orient by looking ahead, behind, to the left, to the right—establishing the four directions of north, south, east and west. This foursquare quality of earth suggests stability. So does soil with its opaque weight, its dark capacity to hold and give birth. Its yin energy nourishes seeds that sprout life from within the mysterious dark womb of matter.

This notion of equating earth with the feminine was commonly held, and not just by the Chinese. Our English word *matter* comes from Latin *mater,* meaning mother. Mother—the *prima materia*, the matrix. This rich heavy mystery of matter provides a passive-seeming container. The receptive vessel for emerging life. Opaque and fecund, the land ties us to it as the mother bearing us all. Sitting steady below the bannering clouds and wheeling stars, it harbors all life.

Eventually the notion of a foursquare earth became elaborated into a mandala that shows the eight compass directions. There is even a triple polarity or "trigram" for each direction, instead of just the binary north and south poles of the West. (Each trigram in the mandala should be read from its inmost line moving outward.) This Chinese octet of triple-deep polarity —or the eight trigrams—was even organized into a primal or old family order, with mother, father and six siblings ranked by birth. This design can be seen in China's ancient art and architecture. Its symmetrical balance of forces became a major motif in the culture, and it is still in use today.

Old Family Mandala

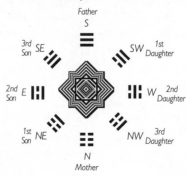

Thus to the ancient Chinese mind, earth embodied the feminine principle of yin. This earthly power bears us forth and then hems us in with its horizons, down with its gravity in a bond that we cannot finally cut, no matter how long we live. We are of this Earth. Yin's wordless wisdom symbolizes a mute truth that holds us stable, grounded, and oriented in the mysterious depth of passing events and our own dark interior. It can be calming, soothing, enriching.

But earth also can engulf us in the overwhelming dark of the unconscious when we are afraid. It will swallow us when we die. Mother Earth births us and then buries us. Dark womb, silent tomb. Dust to dust. Earth carries us back into itself for the nightsea journey toward the great unknown. All this is symbolized by yin. Sometimes yin is embodied in Chinese art as the blue-eyed white tiger, a fabulous beast more unaccountable, wayward and unfathomable than even the secretive house cat. This is earth.

But heaven! Ah! Heaven lofts above the earth, onward, upward, outward—inspiring us with its bright flux and vigorous winds of change and thunderous power and beaming rush of clarity from that emblematic source of enlightenment, the sun.

Heaven symbolized the father to the ancient Chinese, and by association, the masculine principle of yang. Heavenly yang power touches all with its transparent beaming energy as it penetrates every cavity and crevice of earth and inseminates it with the motility of life. It reaches outward to activate, achieve goals, seek clarity, show power. It is symbolized by the dragon, that magical, mystical winged snake. This earth-crawling beast rises to transcend matter.

Thus yang force lofts upward to invisible abstraction, while yin force grounds us in mute material reality. This philosophy was eventually struck into the coin of the realm, and bronze Chinese coins long symbolized the foursquare earth held within the great dome of heaven, the encircled square:

Even now, people around the world still use these old Chinese coins, not as money anymore, but to consult the I Ching. Although I prefer another method myself, I do enjoy looking at them for their shape and cultural significance.

Even Chinese money went holistic. Not *either-or*, but *both-and*, honoring forces that are ever-changing and shifting in a glide through constantly adjusting ratios. And upon this choice hangs what I see as the central philosophical issue of the world's cultures historically. Is any given culture more analog or is it more linear? Traditional, slow-changing cultures like those indigenous to China, Africa, and Australia have tended to be analog, relational, nature-oriented. They've honored the cycling patterns that repeat the eternal verities of nature, again and again with small variations. They resonate to the rhythms of sun and moon and tide and harvest.

But Western culture has driven toward scientific linear progress and so has attempted to divorce itself from the cycling rhythms of "primitive nature." It fondly epitomizes itself symbolically as the heroic conquering male—a giant of strength, of intellect, of prowess, moving forward in huge linear strides . . . yet he is forever being jerked back yet again by that broken record of an analog nag, that specter of dark trouble recurring—usually projected into deadly feminine form as the fickle Lady Luck, or the bitch goddess success, or Adam's tempting Eve, or Socrates' vitriolic wife Xanthippe . . . or updated in the modern media as some witchy woman, perhaps Marlene Dietrich in *The Blue Angel* or Glenn Close in *Fatal Attraction*. Can-do yang on the bright side of life is a sympathetic hero; his unaccountable mate on the yin side is fascinating but frightful, a shadowy woman in a threatening landscape.

More than 2,500 years ago in the West, we began to rise from the *participation mystique* of the group into individual ego awareness. Linearity got the upper hand, and the analog domain became suppressed into the unconscious. Julian Jaynes speculates on this split of the Western mind in *The Origin of Consciousness and the Breakdown of the Bicameral Mind*.

But the ancient Chinese integrated the linear and analog well enough to recognize the cosmic pattern behind chaos theory and to record it in the bifurcating hexagrams of the I Ching. Other so-called "primitive" societies have also shown this integration in various ways—Egyptian, Babylonian, Mayan, Dogon.

The ancient Egyptians, for example, enjoyed analinear fruits for a long period in their early history. It could even be argued that the peculiar landscape which the Egyptians inhabited helped promote their whole-brained approach combining the linear and analog modes. The 800-mile-long green ribbon of the Nile Valley with its annual flood of fertilizing mucky water provided a physical model for merging line with cycle to foster the spiral of evolution. Predictable cyclic renewal urged a goal-seeking linearity to make the most of the Nile Valley's agricultural potential—by developing those quantitative number skills that were involved in hydraulic engineering, in warehousing, in trading. So the rhythm of cycling floods combined with linear engineering, and a great nation arose, thirty dynasties lasting 3500 years, with a long-lived stability that comes from embracing both analog repetition in ongoing process and also the goal-oriented focus of the linear "march toward progress."

But achieving this meld has not been a lasting theme in world cultures. It is as though the globe, like a giant brain, eventually bifurcated into two hemispheres: the East turned predominantly right-brained and analog, while the West opted to go left-brained and linear . . . and seldom did the twain meet. Or if they did, they found each other to be trickily inscrutable or brashly simplistic.

The Western mindset tends to appear simplistic from the subtler holistic perspective of the analog East. The West continually leaps to extremes of *either-or.* It demands to be binary, right or wrong, black or white, hero or villain, male or female, good or bad. It prefers to treat yin and yang as a binary off-on switch, with yin as 0 and yang as 1. Either something is or it isn't. Simple as that.

But yin is not *nothing.* It also is something. The art of carefully doing nothing in order to do something contains within it the analog possibility of walking both paths at once—in the way for instance that 2 doubled and redoubled goes to 8 just as surely as 2 cubed. They are not really the same process—as you can easily see by using another number instead—but with 2s, they reach the same place. This is yin melding both options rather than shunting through gates of *either-or* in a merely binary way.

Yin's acceptance and receptivity, seemingly inert as the earth itself, is the harbinger of creation. Apparent inaction becomes the ground for life. It is honored in the Hindu paradox that says the masculine principle is surrendered assertion, while the feminine principle is creative receptivity. Together, they generate birth. This melding of polarity is a necessary feature of the Tao.

Doing nothing productively is subtle. Yet it is effective. Once in the high reaches of the Swiss Alps, I stumbled upon a cottage with the words *Wu Wei* written on its wooden gate. Nearby a young man was mucking out a goat stable as he listened to Vivaldi on a tape player. Viewing himself as world citizen—as so many Swiss do nowadays—he told me that the gate was doing something just by being there. That's why he named it Wu Wei. Being, it held. In being, it was doing. In talking, I discovered it was his dream to live in wu wei harmony with his extravagantly beautiful natural setting instead of wrestling it into submission. To hold it by being in it.

Creation through wu wei is an analog paradox: yin embraces its opposite in relationship rather than going binary and trying to outdo it with an ultimatum of *either-or*. The proudly yang West has long suppressed the value of its own yin side, calling it illogical, dark, dangerous, ineffectual, engulfing, and unreliable.

To the West, that yin-yang fork looks like a binary 0-1 switch. It sees the two branches as an *either-or* shunt rather than a complementary tension. The 1 means on, active, you're getting information, while the 0 means off, passive, you are *not* getting information.

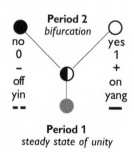

But ironically, even with 0, you are still getting a peculiar kind of data. In binary strings, those non-informational 0s (like these in 100101) also convey information by their very position in the ongoing string. Claude Shannon, the developer of information theory, saw this "information by position" as a weird paradox. Each zero bit of nothing is somehow constraining what comes before or after it in the line to give the whole an added meaning. The 0 enlarges numbers through its very position of relationship—as 10, 100, 1000—without being anything . . . at least to the Western mind. Paradox indeed.

It is interesting to note that even in the 0-1 figures themselves, we can see an implicit female-male quality. The 0 is round, feminine, maintaining a containing space within its hollow center. It rings in nothing and it is also the infinity of doubled nothing as ∞.

But look, on the other hand, at the 1. That sharp masculine 1 is upright, linear, actively declaring its proud existence, defining itself by an erect sharp boundary of "I am!" It is poised to march onward to the next unit, sum, solution.

The linear view sees life in *either-or* scenarios of 0 or 1. But yin is not nothing, just as negative electricity is not nothing. Instead it is complementary—the other necessary pole. You cannot get rid of one pole by pretending it does not exist, just as you cannot get rid of one pole of a magnet by cutting it off. You merely wind up with two magnets, each exhibiting its own pair of poles.

This analog-linear tension lies at the heart of number. It melds both modes into an analinear synthesis whose interplay weaves the underlying co-chaos pattern of the universe.

Workshop 3

Plan a program that combines discussion and experiences. Select book paragraphs to introduce the following topics—or your own.

Opening ritual.

- Ahead of time, ask three people in the group each to prepare a brief example of field-and-ground in ordinary life. Discuss what each demonstration or talk says about human perception. How can we transcend the polarity of yang and yin in everyday life?

- Ask everyone to line up and hold up a forefinger at arm's length. Focus on the finger out there, closing first one eye and then the other. How does your vision shift? What does this say about the coordination between your eyes? This yoga exercises helps to balance your visual perspective. It also has a psyche concomitant.

  Now focus on your finger and then on a picture or spot in the background. Can you hold both in focus? What happens to your comfort level? Now touch your forehead at 3rd eye position. Close your eyes and imagine this touch integrating your insight.

  Discuss your experiences in the group. What do these exercises imply at the psyche level? How can you focus your insight?

- Play a short piece of piano, harp, or percussion music and ask the members to listen to the spaces *between* the notes. What do the spaces do for the notes?

  Meditation: Do you have enough spaces in the music of your life? What kind of music does your life make? Give your piece of life music a title. What group or solo artist could play this if it were transcribed to paper? Should you rewrite your music? How? Ask you unconscious to help you learn and benefit from this insight.

  Discuss your insights in the group.

- Prepare a presentation on the chakras, using quotations from this book or other books. In small groups of 3 or so, discuss which chakra you feel that you are most focused on currently. What issues are involved? What from the chakra station just above can help you transform your problems at this level?

Closing ritual. Five minutes of feedback. Announcements.

Numbers With Heart

The universal web is made of number . . . but it is an oddly relational kind of number. Numbers with heart. They connect and hold the cosmos in synchrony. The loom is patterned chaos.

Patterned chaos is arguably the single most important scientific discovery of the 20th century. You may protest that Einstein's theories of relativity embrace the paradox of space and time unity—as spacetime; or that Heisenberg's Uncertainty Principle confronts the paradox that matter-energy also become a unity—as mattergy.

But consider. Patterned chaos joins spacetime with mattergy. Its patterning holds at every level from micro to mezzo to macro. It can account for all sorts of previously mysterious fluctuations in the very timing and spacing of matter and energy. For example, now we realize that the cycling of epidemics, the rise and fall of cotton bale prices, the formation of hurricane systems, the beating of a heart, the branching of a lung, the rise and fall of a white-blood-cell count, the wax and wane of caribou populations in the Arctic Circle, the turbulent eddying of water in a tidal pool, the drift of smoke upward from an incense stick, the rise and fall of the Nile, the sunspot cycles, the swirling gases that form the Great Red Spot on Jupiter—all of these exhibit the same principle.

What do all these events share? Meshed timing and spacing in matter and energy. It is what creates events. This cosmic pair of pairs sculpts everything into patterns that the modern West calls fractal.

The airways of your lungs are living branching fractals. A neuron reproduces the shape of a Mandelbrot heart—and it's a fractal. A coastline shows the same grainy ruggedness at every level of resolution—fractal. Leaves have fractally-determined shapes. So do snowflakes. Feathers. Blood vessels. Rivers.

Fractals are in all of nature. They can be described by the new holistic science of patterned chaos, which charts patterns that recur on scales large and small in the spacing and timing of matter and energy. The ancient Chinese called it the "nesting boxes of the Tao."

In its very name, this new Western science carries a paradoxical puzzle—pattern and chaos. It is determined because you can predict an overall pattern, but it is also chaotic chance because you cannot specify any exact point of its next manifestation. You can determine its general form but not its exact contents.

Patterned chaos has its own special signature:

- Order in the midst of apparent disorder.
- Cycling that repeats with continual slight variation.
- Scaling that fits one level into another like nesting boxes.
- Universal applicability.

Chaos theory has enabled us to see pattern within apparent random events. With it, we rise to a new level of vision and discover that there is simplicity in this complex flux.

This strange realm first began to be explored during the 1960s, often on makeshift analog computers that charted a peculiar cyclic patterning. Its odd vocabulary of fractals, Julia and Mandelbrot sets, butterfly effects and strange attractors suddenly opened up a new and evocative landscape of visual and verbal imagery.

Chaos science verges on art in its gem-like computer graphics and on poetry in its terminology. It maps a hidden frontier, a new-found-land where linear cause-and-effect confronts the wonder of a holistic pattern beyond ordinary logic. It is so engrossing to scientists, I believe, and it taps into so much creativity precisely because they tend to experience these patterns holistically as well as to analyze them logically.

Benoit Mandelbrot, in his book *The Fractal Geometry of Nature,* makes a statement that is famous among the chaos cognoscenti: "Why is geometry often described as cold and dry? One reason lies in its inability to describe the shape of a cloud, a mountain, a coastline, or a tree. Clouds are not spheres, mountains are not cones, coastlines are not circles, and bark is not smooth, nor does lightning travel in a straight line . . . . Nature exhibits not simply a higher degree but an altogether different level of complexity." Mandelbrot's discovery of fractals opened a scientific door to make nature accessible in a new way.

These fractal patterns are incredibly huge, intricate mandalas hidden in number itself. And mandalas are powerful. Psychology knows that the mandala shape acts as an organizing and healing impetus, and when a patient begins to be preoccupied with mandalic organization, to draw it or to seek it out in a rose window or in the still center of a maze or perched in the hub-like view from the Arc de Triomphe, then psychological improvement is on its way.

How provocative it is to realize that this mandala of number, this holistic pattern of the fractal, has been discovered just when nature is sorely blighted by a mechanistic, GNP-oriented culture that has lost connection with its environment and its soul. Is it possible that nature is seeking to heal itself from within by revealing this fractal geometry now in our global society? Can it be that sick nature is curing itself by this manifestation of vast hidden beauty revealed to those who have most destroyed it—us humans? And more specifically, to that coolly objective and linear part of ourselves that is enshrined in science?

The baroque mandala of the Mandelbrot set has been called the most complex and universal number relationship in mathematics. It describes designs of infinite depth, using fractal geometry. At its core lies the Mandelbrot heart. This dark heart was discovered by Mandelbrot in 1979 when he began to chart a pattern of awesome beauty within number itself. This heart appears in many different formats. It can be seen on the cover of this book underneath the title. It is also visible within the Chinese vase of the colophon. A stylized version is used for the bullets in this book . . . for example on the opposite page. Following are some variations of the Mandelbrot heart that have been taken from four different computer screens.

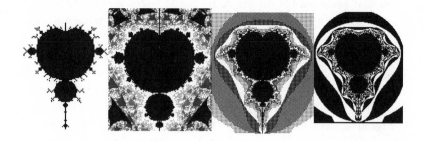

Mandelbrot Hearts

These versions of the Mandelbrot set look like valentines edged in frothy lace. Each dark heart lies at the intersection of real and imaginary number, on a sheet of paper or a computer screen. A simple equation is the source. When numbers are plugged into the equation, they define a central black area of all the numbers that do not go to infinity, but instead are either trapped in a repeating loop or meander chaotically. Around the dark heart lies a lacy fringe holding all the Julia sets. They weave a baroque contrapuntal border of relationship at the edge of numbers starting their slow acceleration to infinity. Everything beyond this lacy border drives to infinity in an escalating rush of increasing number. The cover of this book shows part of a Julia set. It is colored number.

The sheer beauty in the Mandelbrot mandala grips us as much as the elegant simplicity of its "nonlinear" equation surprises us. It is mesmerizing. Much of the appeal of patterned chaos lies in the visceral thrill of its profound beauty. Benoit Mandelbrot and others during recent years have made many converts to the study of fractals simply by carrying along on their lectures a slide tray showing this awesome intricate jewelry of nature.

Furthermore, here is a beauty that the scientist can interact with rather than merely view from a logical remove. The West has often sought to rape the beauty of nature precisely because it is viewed at a logical remove, defined as something passive, malleable, feminine, yin, conquerable. Science has taken a tacitly yang stance that tended to dominate nature by logic's divide-and-conquer tactics. It drove toward a climaxing solution rather than relating with nature as a partner in an ongoing creative process.

But the scientist can explore the beauty of patterned chaos only by interacting with it. On the computer screen, for example, the explorer discovers evolving networks of number in holistic connection. Jewel-like patterns shift their shadings of color in the shallows and depths of number like continental shelving along a coastline. The fractal patterns show a rich detail in their subtly ever-new proliferation of the Julia sets. Sea Horse Valleys appear. Lions. Dragons. Imagination responds to the infinite depth of rich jewelry in color and form, much as treasure hoards of rubies, emeralds, and sapphires in gold filigree filled the folk tales of ancient times.

It can even be frightening to discover how subjectively one responds to this panorama of natural majesty. A. K. Dewdney in *Scientific American* describes viewing it on a computer whose zoom lens ability can take a section of detail and keep magnifying it to endless resolves of beauty so that " . . . a riot of organic-looking tendrils and curlicues sweeps out in whorls and rows. Magnifying a curlicue reveals yet another scene; it is made up of pairs of whorls joined by bridges of filigree . . . . As the zoom continues, such objects seem to reappear, but a closer look always turns up differences. Things go on in this way forever, infinitely various and frighteningly lovely." The Mandelbrot set iterates its elegant shapes of nesting infinity to hold for the scientist a visceral impact that is more powerful than mere logic can acknowledge.

Science responded to this mental and physical beauty in fractals by turning to explore nature and suddenly finding fractals everywhere. The basic principles of chaos theory can be applied across disciplines with an assurance seen in the past only by the blanket, unexamined dogma of some religion or dictator. Yet its broad application bears fruit simply because the theory holds true so widely, so provably, so productively within bountiful nature itself.

The universal import of this odd new science is even tending to heal the ever-dividing rift of disciplines that fragmented the past centuries into splinter groups specializing ever more narrowly in describing more and more about less and less. But now cardiac specialists, stockbrokers, psychologists, and turbulence physicists can all meet in joint conference on chaos patterning. A recent conference held in Florida sponsored papers on chaos patterns in physiology, biophysics, gestalt psychology, chemical systems, mathematics, physics, communication theory and linguistics. Such unity within diversity was unheard of not so long ago. Here indeed is interdisciplinary reunion. Patterned chaos can find shared meaning in areas as wide-ranging as plasma physics, genetic coding, perceptual psychology, coffee prices, weather reports, the body's living ductwork of blood vessels, airways, and nerve impulses.

Let's take a simple example of the chaos dynamic in action. A friend brings a gift over to your place: it's a single seed with a pot of soil to plant it in. Your friend says, "I knew you wanted a house plant. Well, this is a Silly Centimeter vine. This seed here is 1 centimeter long. After you plant it, the next day it will come up. Each day the vine grows its current length squared plus one silly centimeter."

Together you plant the seed and look at the pot. No growth is showing yet—it's at ground zero. But you know that the Silly Centimeter seed is inside there. You set the pot over in the corner by the window and walk away.

The next day you come back and the plant is up. How cute. It's 1 centimeter tall. But well, that's less than the length of your little fingernail. And you did want a nice large plant for this corner of the living room. Some refreshing greenery. You just hope the plant won't stay too small.

The next day you come back and the vine is 2 centimeters tall. How nice. It seems to be doing all right in this location. The next day you come back and it is 5 centimeters tall. Good. Maybe you'd better run a string down to it from the ceiling, so that it will have something to climb on. You put up a string.

The next day you come back and your Silly Centimeter vine is suddenly, abruptly 26 centimeters tall! Nearly a foot tall. Well. It's going to look good in that corner, a nice spot of green. The next day you come back and it's—what!—677 centimeters tall! Over 22 feet! It's looping around in the ceiling corner above the window. Humm, maybe you'd better train it to grow around the window frame if it's going to be this long!

The next day you come back and your Silly Centimeter vine is 458,330 centimeters long! That stretches over 4 ½ kilometers, or close to three miles! It's filling the living room with closely packed green vine and winding down the hall. You must have somehow gotten hold of Jack's beanstalk! Looks like you're going to have to find someone to come in and destroy this monstrous thing and haul it away. But whom can you call? You've heard of Jack the Giant Killer, but who hires out to do away with giant beanstalks? A neighbor suggests a tree removal service.

You phone the Jiffy Tree Removal Service and they come over to your place the very next day, but by that time your Silly Centimeter vine has grown to the astonishing length of 21 kilometers! Over 13 miles long! Here it's been only one week since you planted that Silly Centimeter seed and already it's driven you out with its jungle of green vine throughout every corner of your home.

Now. The whole reason that this Silly Centimeter vine got so big, of course, is that it is squaring its growth plus that one silly centimeter added each day—and using this total as the basis for the next day's growth. It is cycling its daily result into the squaring that makes the next day's growth.

Benoit Mandelbrot used this very cycling process when he explored the analinear equation that defines the Mandelbrot set in the math of chaos. It is a very simple equation, about as simple as you can get and still mix a discrete lumpy constant with a sliding analog cycle. Each cycle is called an iteration, because its cycling process says, "Play it again, Sam. With a variation."

The Mandelbrot formula acts just like our Silly Centimeter vine. It starts with a squared 0, and so it begins at ground zero. Then it multiplies that nothing—the 0—by itself to start the process going, and also it adds in a lumpy constant unit of 1 (just like our Silly Centimeter vine added in the constant lump of 1 centimeter). That first cycle's squaring gives nothing, because 0 squared is still just 0, but recall that we also add in the solitary constant of 1. Nothing plus 1 gives 1. That completes the first day's growth. It is the first cycling, the first iteration.

To do it again, you take that first cycle of growth, the answer of 1, and square it. But 1^2 is still just 1—and again you add in the lumpy constant you started with—that silly 1 centimeter which was our vine's constant. Now 1 plus 1 makes a total of 2. And with each cycle, you keep doing this. Keep recycling your answer, iterating this process to inflate the ballooning ratio of squared number and each time you add in the lump of that discrete constant 1 to see how fast your rising tide of number screams off to infinity.

Simple enough, right? But it makes a Silly Centimeter vine grow wildly. There is no final product to it, no sum, only an arbitrary jumping off point where we choose to stop. The process itself runs to infinity. Somewhere that Silly Centimeter vine is still curling upward. Each new bigger answer feeds itself back into the squaring process and keeps the dynamic escalating. The squaring number provides the smoothly rising volume of analog cycling, but the continual addition of that constant lump of 1 throws in the discrete linear number. Voila! A "nonlinear" equation . . . or really, analinear, for it combines both the analog cycle and the chunky unitized line.

This equation describes more than escalating growth, though. Hidden in it is the secret of patterned chaos, the Mandelbrot heart. This equation tells how to grow the Silly Centimeter vine and also the Mandelbrot heart of determined chaos. Computation is what makes the Mandelbrot equation go. Fortunately it can be calculated in the quick-time of computers that process number so much faster than people can . . . or even want to, because it's so incredibly boring. Monotonous. In fact this equation is what is called a monotonic process because it just keeps on growing monotonously.

So the trick is to get some significant variation into its growth pattern. That Silly Centimeter vine had only one effective dimension to it, the line of growth pushing onward. It was just an extending line. Of course a real vine would have thickness too, but that didn't matter in our story—length was the issue.

But what if we plant ten thousand seeds in a container a foot square and they all have different growing instructions? Ten thousand different growth-rate lines. What if we tell them to grow upward, yes, but do it at different speeds and in different directions? Give each seed its individual "Simon Sez" instructions, maybe to "Take one giant step south and then two baby steps west and then go . . . ." So they all start growing in their individual rates. Those on the outer edge are zooming up fast, very fast indeed. Stratospheric!

We get in our spaceship and from our orbit way above, we start taking pictures of these myriad vines coming up. The different growth rates are color-coded. As we photograph them from above, we begin to notice an overall picture that these vines are forming together. It looks like a complicated weaving pattern.

Good heavens! The design in the middle—that slowest growing area—it's the Mandelbrot heart! And around its sides, in the highly variable growth rates, are the colorful fringes of the Julia sets! Look at that shelving off the outer banks of the heart—it is number growing lickety-split to infinity!

That's also what grows straight out at you from the computer screen. Ballooning number. The many initial number points that are scattered over the screen are even called seeds. The seeds you plant at the various points have number addresses on the screen. The reason this equation graphs out so prettily and shows its mandala heart is that each seed on the screen has its own address with its own growth instructions.

Mandelbrot didn't just grow a single line of number zooming straight up as we did with our Silly Centimeter vine. No, first he gridded the flat plane of the computer screen with coordinates. Then he seeded this map with dots, each with its own growth instructions. This wove many vines into a symmetrical jungle rising off the screen. As each seed grew, he watched to see how fast its particular rate of growth leapt toward infinity, using the series of ever-extendable decimal numbers lying on the map between 0 and 2—numbers real and imaginary, plus and minus. Finally he color-coded these different rates to look at the overall pattern.

In the center of the screen was the dark heart limned in by its surrounding number frisking off in an elaborate dance toward infinity. Other tiny hearts were embedded in the fringes, actually an infinity of them, and all of them were connected by a well-nigh invisible filament. This heart kept repeating in infinitely nesting sizes, however incredibly fine Mandelbrot gridded the screen—not exactly the same heart, but almost. A cardioid shape, scientists like to call it—for who in cool science wants to refer to lacy valentines?

Of course, the trick in getting the different growth rates was to give each seed not just a simple "grow up" message," but also a "grow around" message to map it over to new locations and so put a weaving torque into it. To get this effect, Mandelbrot used for seed not just number, but complex number. It gives each seed those twisty "grow around" instructions so that the growth is not a monotonous shooting straight up. Instead it can weave up and around and down. By means of this complex number weave, the brilliant mandalas formed on the computer screen.

What Mandelbrot basically did was clock each location's escape velocity and compare the rates. Watching how fast things change means that you are essentially watching timing in spacing. The chaos dynamic weaves patterns in the timing of spacing. James Gleick described this relationship in *Chaos: Making a New Science:*

"But when a geometer iterates an equation instead of solving it, the equation becomes a process instead of a description, dynamic instead of static. When a number goes into the equation, a new number comes out; the new number goes in, and so on, points hopping from place to place . . . ." These points make patterns that are predictable in form but not in the placement of each specific dot. You must relate the points to each other; you must fill in the dots to get the big picture. Seeing the dots make the pattern is what creates the Mandelbrot set. It is what finally embosses that deep design into the center of the escalating numbers. Fortuitously, this feature also makes the color spectrum very effective in showing its patterning.

Color is seen as quality by our consciousness. Of course the light wave lengths can be quantified by scientific techniques, but that does not impart the quality message that color gives to our attention. We experience color as the flux of analog relationship. A color changes its impact according to its surroundings. Red with pink has quite a different effect on the eye from red with green. This book cover fills in the dots and color-codes them to show the design.

Consider these whorlings of the lacy Julia sets. Even in black and white, they are marvelous. Continual shifts in their boundaries are caused by fractal attractors. In this intricate whirling, no border between any two attractors ever forms without at least a third attractor also insinuating itself, so that there is no simple *either-or* boundary. So boundaries keep shifting. Talk about being betwixt and between!

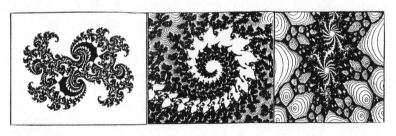

| Escher's Julia | Newton's Julia | Nautilus Shell |

It is easy to see boundary shifts in physical matter. When matter shifts its physical state—for instance when H_2O changes from ice into water, or from water into steam—this boundary shift is called a phase change. The shift-over moment is evident way down at the atomic level, as atoms alter their ways of relating to each other.

How come? Well, atoms contain built-in polarity. Sometimes these tiny magnets are scrambled in random alignment, like in water—but sometimes they line up their individual poles to exert a unified force, like in magnetized iron. Physicists Yang and Lee even showed how this magnetized polarity has a fractal nature!

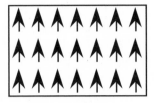

polarized magnets   unpolarized magnets

The Mandelbrot heart hides everywhere, even in atoms. Rules for chaotic systems can vary from one another and still arrive at the same central heart. Peitgen and Richter point out in their superbly illustrated book, *The Beauty of Fractals,* that "The details of the rule are not essential because we will see that different rules may lead to the same Mandelbrot set." These two men sound much like explorers suddenly stumbling upon legendary Lake Victoria in Africa as they recount coming abruptly upon the familiar shape of the Mandelbrot heart deep in what appeared to be random number. While investigating the phase boundaries for magnetism, they stumble through number wilds "in the neighborhood of the black region . . . bewildering . . . progressively tangled . . . with a surprise: the well-known Mandelbrot figure appears, its identity with the original Mandelbrot set is astounding." These rigorously trained and careful German scientists, one a mathematician and the other a physicist, declare simply, "Perhaps we should believe in magic."

This is the seeming magic of chaos patterning. Indeed, it is energizing to feel the thrill of magic again after so many years of cool rationality in the linear halls of Newtonian science. For this magical mandala is not just a flashy gimmick or wordless wonder. It has heft and weight and is supremely useful. It has begun to heal the rifts caused by the logic-chopping, divisive compartmentalizing of Cartesian science. Human knowledge is unifying again through that analog twist that is given to linear number in patterned chaos.

Shades of Taoism! Taoism with its nesting boxes of events in patterns beyond logic! Shades of synchronicity, of what Carl Jung called the acausal connecting principle, of spacetime and mattergy meshing to reveal hidden meaning in random-seeming events.

This connectivity is not truly acausal. It *does* have a cause. Chaos patterning. Fill in the event-dots of your days to trace out the hidden patterns in the spacing and timing of your life. Patterns deep in the welter of data show the psychic energy that Freud called complexes and Jung called archetypes. Your unconscious life will cycle in huge and tiny self-similar patterns even as your linear logic is busy marching along in conscious cause-and-effect intent. Together they weave the pattern of your days. Your unique identity asserts its main motifs in large and tiny ways again and again, never quite the same and always farther along the time line of your existence.

Psychology knows that a complex can shape your behavior in the midst of apparent random chance. It puts repeating motifs in your life with continual slight variations—in the lingo of bestselling self-help books, you may be an eternal Peter Pan, for instance, or you may have a Cinderella Complex. To use motifs from 2,500 years ago, you may exhibit an Oedipus or Electra complex. Whatever its name, your complex will constellate events on scales big and small, locking you into iterations with slight, ever-new variations.

A bad habit or negative complex can cycle on and on even as your logic is fighting it. To recast that analog pattern is not easy. It takes effort and perspective. It means entering that liminal space where change is possible. And that scares the entrenched ego. It resists the phase changes of ego death and rebirth.

You recall that the Mandelbrot set has borderline conditions or transitional moments called phase changes. In a psyche's phase change, you just don't know where to stand psychologically. For instance, you're losing your smoking habit/job/home/best friend/ lover/child/good health, and don't know how or where or when or if any relief is coming. You lose the old footing. The roof is falling in. Tornado! Earthquake! No secure boundary demarks what is safe, what is not. Events feel chaotic. You whirl in possibilities that pull you momentarily this way, that, the other. Habits break. It's not *either-or*, but instead, far more complex. And then finally, events move you on through that furiously fragmented borderland into a more settled phase . . . until eventually, it alters again.

The borderland is an undecided domain, but by maintaining your overview, whether it be on the computer screen tracing out a Julia set's fantastical twists or in your psyche's gyrations, you can turn bewilderment into patterned meaning. Boundary conditions occur in the fluid psyche as well as in the escalating numbers of the Mandelbrot set, and being lost in the whirl of the archetypes is the psychic equivalent of being pulled about by the fractal attractors of a Julia set. The only way out of its grip is not through straight-and-narrow linear logic, nor through mindless iteration, but using an analinear transcendence that combines the two modes and lifts you up to see the overall pattern and redefine your position in it.

Think about this. Better yet, feel it. Each day iterates your life in a new variation on a recurrent pattern, yet it never quite retraces the old path. Your thinking and feeling regard this next development and feed that back into the equation, which subtly alters your reality. Each day feeds back into the mix and subtly redefines your life's daily dynamics . . . much as with that silly centimeter vine.

The equation of your life becomes ongoing process, not a solution. The constant in this equation is you. The cycling iteration is daily events. A hidden pattern slowly emerges from this interplay between the constant of you and the sliding data of events. You inhabit a moving design that you can't even see very well because of the limitations of being finite. It's hard to find that lofty spaceship to get up high enough and take an overview of your own growth pattern.

Mostly we live deep in the weave. But the sudden occasional emergence to an overview is amazing. You stumble upon a stunning vista of gut-wrenching pattern, and the body responds to its shocking beauty as forcefully as the mind does, but even quicker. Reaction is part and parcel of the analog. Something resonates in you, it rings your chimes. Turns you on, tunes you in. You hear life's vibes. Its subliminal music rises to audibility and resonates all about you, sweeping you along in its majestic, poignant, witty, glorious, awful swells of beauty.

The very frustrating thing about this analog resonance is that it is so hard to evaluate in a merely linear, logical way. Even the senses work by comparison and association, inviting a subjective, relational commentary: "That noise is too loud!" "Oh, I like that color." "Tastes like chocolate, my favorite!" "It's gardenia . . . oh, that scent takes me right back to my high school prom corsage." Sound, sight, touch, taste, smell—they all trigger analog resonance. You experience them emotionally in a qualitative response. The peculiar fact is that your reaction can happen even before the sensation is registered in your thinking brain—that is, in the brain cortex. In other words, emotion can come before conscious thought.

People react before they even know why. Joseph LeDoux of the Center for Neural Science at New York University has found that emotion registers not up in the more evolved cortex of the thinking brain, but way down in the lower, more primitive brain stem of the amygdala, that almond-shaped spot where dream emotions register. Amygdala is Greek for *almond*. (Aptly enough, Rollo Silver publishes a fractal newsletter called "Amygdala," but in honor of Benoit Mandelbrot, whose last name means *almond + bread* in German.)

The psyche is a living system in continual flux, as surely shaped by the dynamics of chaos patterning as your lungs, feet, and hair. Its complexes, archetypes, behaviors are dynamic patterns that are visible only through motion and emotion. And that's devilishly hard to chart, precisely because it is analinear. Studying a holistic living system like the psyche demands something more than the old Newtonian lab techniques of dissecting, weighing, and storing data. It just doesn't work to chop up living nature into discrete segments. We kill it. A mind is more than a pickled brain on a shelf. Trees become chunks of wood, a dog becomes organs on a table, the global ecosystem of Gaia becomes two inches of rain in the gauge outside the back door. In such logic-chopping measurement, we disregard the larger picture that makes a holistic system more than the sum of its parts. It is a web of life in vitally balancing factors, not a chance agglomeration of dead units stuck together.

Of course there have been lots of ways to study death. Nature can be divided into various bowls and dishes and vials. It works well for exploring the "nonliving" world full of mechanical action. For a long while we learned much by chopping nature apart and wrapping our thoughts around it to form packets of formulas and equations in linear rows that weigh out a final product as quantity.

Yet more and more, we hunger for universal meaning. We sicken without knowing why, bereft of our kinetic, unspoken relatedness. In some unconscious realm we have begun to fret and know what we have lost. While erecting those smooth lines of logic toward the heights of infinite perfection and finding that unreachable, we have also lost some rough Eden of connective wholeness.

But 20th-century science started moving back toward center. It described a growing mandala of cosmic connection that lies deeper than Western science had known before, that was seen previously only by mystics and sages and artists who experienced it but could not explain this felt connection in logical terms.

This mandala of cosmic connection holds the secret of change. Here is why the ancient Chinese sage-scientists studied change, and indeed named their findings the I Ching or Book of Changes. Modern chaos patterning sounds oddly like those benighted old days when ancient scholars saw the cosmos as nesting boxes and said you could tap into any level of the pattern by meditation. By giving the pattern your sincere attention, you could change its energy and thus alter your relationship to it. They even developed the I Ching to describe the varying quality in the flow of events, without bothering to explain exactly how it happened in a *quantitative* way.

Western science is mostly an elaborate collection of hows, but Eastern science was mostly a collection of whys. The East talked of analog qualities, not cause-and-effect quantities. It conceived of nature in a holistic and relational way, while the West described it "objectively" using a separate, distancing point of view, discretely linear, as a mechanistic and ultimately meaningless flux of events. To Western logic, ultimate meaning became a silly delusion. Steven Weinberg's remark in *The First Three Minutes* summed it up: "The more we understand the universe, the more pointless it seems."

But the watercourse way of the Tao has a gravity gradient toward meaning, even if one cannot understand the topology of a specific event. In the landscape of the soul, the stream of the Tao flows into a sea of meaning. Struggling upstream against the way of the Tao is counterproductive and wearisome, but understanding and merging with its purpose is reviving and reorienting. Following it, you discover who you really are and the task in life that you were created for. By enacting your own task in the pattern, with its unique dynamic and significance, your life takes on transcendent meaning.

From culture to culture and age to age, we deposit our psychic projections about natural power at the forming edge of knowledge. For the 20th century, it was science—in the quest for black holes and DNA structure and uniting the four forces. The numinous resides just beyond the limits of our most honored scientific study. Why? Because we seek to find and honor truth in the not-yet known. At this boundary we glimpse the fragile, ever-changing face of god.

WORKSHOP 4

Plan a program that combines discussion and experiences. Select book paragraphs to introduce the following topics—or your own. Opening ritual.

❣ Ahead of time, ask if anyone has slides or a video of chaos patterning. (See ordering information at back.) If so, use some group time to show these amazing jewels of number order.

❣ Ahead of time, ask each member to bring a small object that has personal symbolic meaning. At the session, use a small table as an altar and ask each member to place on it the object, describing why that object matters so much. When all the objects are on the table, light a candle or sage bundle to honor the meaning behind this mandala of matter sitting on the table.

Meditation: Play music while meditating to see meaning behind the shifting matter of your life. If desired, read aloud paragraphs from the last pages of this chapter. Ask the Tao to shape this flow of matter through your days into significance, service, and joy.

❣ Bring a pad of newsprint and marking pens so that everyone can draw while sitting at tables or on the floor. The task—draw your life. Put in the four earmarks of patterned chaos: order in apparent disorder, cycling with variation, scaling levels, and global applicability. Do not try to plan your whole picture ahead—just begin and let it grow. Then in small groups discuss your picture, explaining what the images signify. What does it tell you about your life? In the large group, let each person share a sentence or two summary of what you see about your life pattern. What trends and habits do you like, what do you want to change?

❣ Habit Swap: Pass out 2 cards each. Ask members to recall a habit of theirs that is beneficial and write it on Card 1. Take up all Card 1s. Next, write on Card 2 a habit that is not beneficial. Take up all Card 2s. Let each person draw from the Card 1s at random. Now let each read a card aloud, claiming that useful habit as their own. Enjoy it; rejoice in it. Next, let each person draw from the Card 2s at random. Let each read a card aloud, claiming that difficult habit and deciding how to befriend it to work with it for change.

Closing ritual. Five minutes of feedback. Announcements.

Chaos Rhythm at the Core

Rhythm beats deep in the heart of chaos patterning. Its music sings us into existence, body and soul. Following even just a couple of the major motifs in chaos theory will reveal this symphonic union between modern science and ancient mysticism.

The first major motif is bifurcation. It means splitting in two. Forking. Two roads diverge in a yellow wood. Two friends head in different directions. You take the high road and I'll take the low road. He lives on the sunny side of the mountain; she lives on the shady side of the hill. In fact, yang originally meant the sunny side of a hill, while yin meant its shady side. As the brightly lighted foreground took centerstage, the shadow retreated away from attention into the background.

Sometimes a behavior will fork into two behaviors that alternate back and forth. Let's say you look out the window and see a flag draped in lank folds on the pole. It's not moving. But later when you look again, the wind has come up, and hey, now the flag is flapping. Then it sort of collapses into a buckling motion. Back to a flap. Then it buckles again. You watch to see if there's any regular rhythm. Yes, right now it's alternating flaps with buckles.

Thus it has forked from a steady state into two alternating states. For the moment at least, it dances in a pattern of either flap or buckle. So you decide to draw a little tree graph to represent this forked behavior.

First, you draw a single line straight up to represent the original steady state of the unmoving flag. You call it Period 1, meaning just the trunk itself. Then higher up, you fork into two branches—the alternating flaps and buckles. They form Period 2.

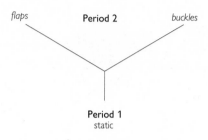

flaps   **Period 2**   *buckles*

Period 1
static

Flag Motions

At any time the flag could start flapping briskly with no alternating buckles, or just buckle and buckle again, or it might drop back into its old static condition of lying draped lankly on the pole. But not right now. Instead, it develops an even more intricate little dance. The flap on the left further splits into a sharp crack and alternatively, a hollow boom. So you draw a new, higher fork on the left to show the cracks and booms. Meanwhile, on the other side, that buckle action is also forking—into either a twist around the pole or a shudder. You draw it on the right as two new higher branches.

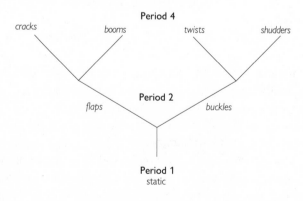

Period 4
cracks   *booms*   *twists*   *shudders*

Period 2
flaps   *buckles*

Period 1
static

Flag Motions

This gives Period 4 . . . because it is four branches wide. All four motions are now interplaying in a complicated beat of crack, boom, twist, and shudder.

To tell the truth, this flag dance is a simplified event. Your eye is not a stop-action camera, and it all would happen so quickly on a flagpole that you couldn't easily discern these distinct motions. However, for the sake of our bifurcation model, let's double the action once more by branching upward into another level of this busy little dance atop the flag pole. Here comes Period 8 with its snaps, beats, bams, pops, hums, crinkles, shivers and shakes. Unity has forked into 2, then 4, and now finally 8 states.

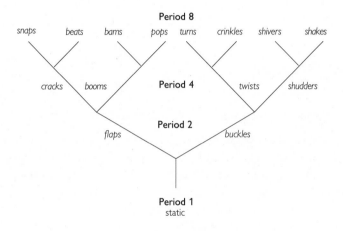

Flag Motions

This tree doubles its branches at each new level of growth. That's why bifurcation is called period doubling. And it's also possible for one side to fork upward but not the other, so that you might wind up with an odd number of branches, giving a Period 3, or Period 41, and so on in a horizontal "window" view across the branches.

Following is a bifurcation tree with an inset showing a window where only three periods of branching appear—in other words, it has a Period 3 window. This Period 3 window will be important to the theory of co-chaos later on.

Bifurcation Tree

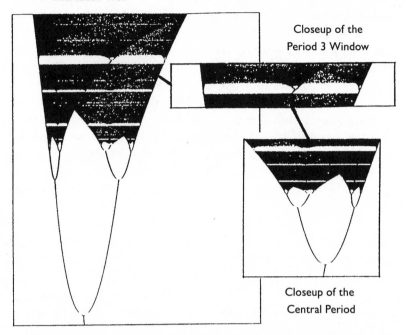

Closeup of the
Period 3 Window

Closeup of the
Central Period

Bifurcation Tree with Inserts of Period 3 Window & Central Period

Notice how this tree just keeps on forking upward, until the flag events all quickly degenerate into random busy motion whipping about on the flagpole. Chaos, to all appearances. The growing tree no longer shows the clean black branches of bifurcating behavior. Instead it jams into the obscuring twigs of a continuous black welter of chaotic motion, infinite cascading bifurcations. Usually we live in just such a forest of events, where we cannot see the integral trees for their obscuring twigs of random-appearing details.

This bifurcation can keep on cascading in complexity, sure, but it may even revert to a simpler level of complexity. In fact, if just three branches exist, we've still got relationship between relationships. It is potent relationship, in fact, the volatile energy of 2 to 1 in various shifting balances. The infamous triangle.

If the twigs ever do clear away momentarily to show just three branches evident in a Period 3 window, then the dynamic carries within itself the kernel of not just random chaos, but *patterned* chaos. In fact, if you're not scientific-minded, here's all you need to remember—this Period 3 window. It's the whole key to chaos theory.

It holds a program able to go to the heights of twiggy obscurity and still return to its lower, less complex pattern . . . and also go back up again. Notice that this insert of the Period 3 window has a Central Period that starts branching again back to twiggy welter. It is this Period 3 window that certifies these random-seeming events have developed a behavior that will keep attracting reinforcement to it. It becomes "matter with a will of its own," in Ilya Prigogine's phrase.

Only a stop-action camera could discover the wonder of wind flapping a flag into chaos patterns. Chaos dynamics is a bit easier to study in liquids, though, which move more slowly than air. It is easier still to study in solids, which move even slower yet. Take the solid flesh of animals, for instance. Life exhibits rhythms that feed an ever-increasing literature in chaos theory. Fish, insects, plants, birds, mammals—all exhibit rhythmic, bifurcating patterns in their body structure . . . and also in their life cycles.

Since a life span is longer than the crack or pop of a flag on a pole, it is easier to chart. Here is life at its coarsest resolution, the birth/death rate of a population. Biologists have charted statistics on flies, sharks, bobcats, deer, rats—you name it. Yeast, for instance, will double in dough every 20 minutes at 33° centigrade, but only as long as it has food, the flour's glucose. When it has no more to eat, it begins dying . . . and so most bread recipes will add a bit of honey or sugar. You can even keep a sourdough starter alive in the refrigerator by suppressing its appetite with a cooler temperature and also giving it more flour once in awhile.

Bifurcation may be graphed downward into a well instead of up into a tree. When a population cycle is completely stable, it has only one attractor point, that static state where nothing new happens. This single fixed point of Period 1 sits in a well, unable to go anywhere. But in Period 2, the population cycle bifurcates into two complementary rhythms. We have now two wells with an attractor point sitting in each.

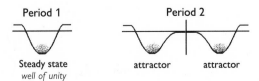

Often an animal or fish or bird or bug population cycle develops a regular oscillating pattern with two attractor points sitting in two wells (the 2-beat limit cycle). Or it may bifurcate into four or five attractor points and then rush onward into apparently random chaos . . . but then eventually it moves back into a stable pattern with a 4-beat or a 2-beat or even a monotonous steady state in a static well where the number born consistently equals the number dying. Zero population growth.

But this chaos process is bigger than living organisms; it is typical of events in general. Leon Glass and Michael Mackey point out that bifurcation is the universal trait of patterned chaos. Whenever it occurs, patterned chaos is suspected, and it comes at an accelerating rate that is predictable in every chaotic process. From all this, and much more, James Yorke and Tien-Yien Li proved that it takes only *one* Period 3 window to give the essential conditions for determined chaos. In a short article packed full of heavy math, their title is by far the liveliest writing. Fortunately, the title in itself—"Period Three Implies Chaos"—tells us what we need to know. Once you get a Period 3 window, you've reached the point of no return, and it's patterned chaos.

Our flag tree below develops into a Period 8 window of timed events. Oddly enough, geometry does likewise. This tree bifurcates in time, but geometry does it in space.

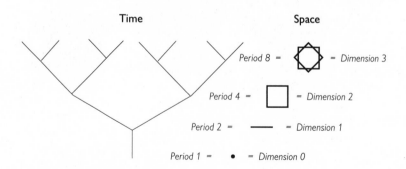

Time and Space show Bifurcation

In space, a point is dimensionless, static, unmoving—like our tree trunk in the static well of Period 1. But a line oscillates back and forth between its two end points to reveal a first dimension called width—and voila, it's Period 2. Next, a square cycles among its four corners in the second dimension called height . . . and it's Period 4. Finally, in Period 8, a cube cycles through its 8 corners to sketch in that third dimension of depth. These three dimensions are formed by polarized spatial bifurcation. Bifurcation is truly universal.

A second major motif of chaos theory is the fractal. *Fractal.* An odd word and odder behavior. It is a descendent of bifurcation. A fractal can repeat itself with variation on scales large and small, in space or time or matter or energy. Or all of them together.

A famous fractal is the Koch snowflake, described in 1904 by Helge von Koch. To make it, you take a triangle and "grow" a new triangle onto the back of each side, just like the old triangle, but only 1/3 its size. Thus a side that is 3/3 long is replaced by a new side that is 4/3 long. You can keep piggybacking triangles on top of triangles to create a rough serrated disk full of interior stars. But way before Koch described it mathematically, this figure was in religion, for instance, as the 6-pointed Star of David.

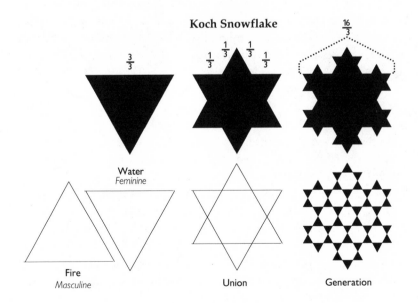

Koch Snowflake

Water
Feminine

Fire
Masculine

Union

Generation

Yantra

The Koch snowflake is also common in the Far East. It's seen for example in this Hindu yantra drawn in Nepal around 1750. A yantra is an image to help you meditate on the rhythmic timing of energy, just as a mandala helps you meditate on the harmonious spacing of matter. In the center of this yantra, six Koch snowflakes form a serrated disk with a perimeter of eighteen points. That seventh snowflake in the center suggests the final bindu point of union with god. Around this central disk, the eight black and white doubled petals suggest the yin and yang of the eight doubled trigrams that form the sixty-four hexagrams of the I Ching. On the exterior of the design, the four T gates open this sacred temenos of holy order to the ordinary world outside.

You can keep growing new, tinier triangles on the backs of old on your Koch snowflake. Again and again, as long as you like, even to infinity. Finally, at infinity, you will have an infinitely rough shape that is smaller than your hand. This little snowflake sitting your hand with its tiny burring of points has an infinite length around its outer edge, a line completely full of points, and furthermore, each of the points has a sharp kink in it!

A single kink on this snowflake has been compared to a rough peninsula on a coastline. In the larger view, a coastline is just a line. Water cuts an arbitrary slice right across whatever comes down or up to meet it—cliffs, sand, boulders. You get a rough and bumpy profile wherever it meets the land. A silhouette of their connection.

And this profile has the same general appearance on every scale, large or small. If you photograph two hundred miles of coastline from thirty thousand feet up, you get a certain rugged effect that is also evident when you walk along and photograph only ten feet of it close up. The small profile will carry the same rough and ready fractal proportions as the big version, just in a different size. It reveals the self-similar scaling of fractals that is so evident in nature. Thus the Koch model is intriguing because, like the world nestled in William Blake's palm in a grain of sand, it holds infinity in its rough and tiny microcosm.

Here is another fractal image. It is a square snowflake. And like the Koch snowflake, it also shows three orders of development. This time, though, you are replacing each side with one *twice* as long—going from 4/4 length to 8/4 length. If you keep doing this, at each higher order of evolution, you're making the margin twice as long. Of course it could go on forever, and you'd wind up with a strange jagged shard whose border is infinitely long yet you could put the whole thing into your pocket.

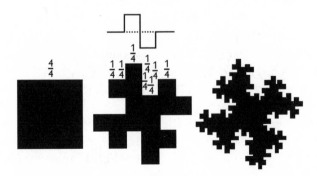

Square Snowflake

These odd snowflakes are fractals. Their margins never cease making jogs and can elaborate in self-similar scaling to infinity. So it's no surprise that they also have peculiar needs when it comes to measuring their dimensions. They ask for the special license to have "fractional dimension"—to be a fractal, in other words.

Fractional dimension? It is a mind bender. Common sense says that surely there can only be whole dimensions—width, height, and depth. Okay, go ahead and add Einstein's fourth dimension of time. But still, that doesn't add up to a fractional total.

Trouble is, mathematicians don't really mean "dimension" here in the ordinary sense. It's rather like calling an African lion and a house tabby both cats. Sure, they do have shared traits and a certain common heritage. But Fluffy does not really describe the color, size, habitat, and nature of Leo, lord of the jungle. It's a whole other breed of cat. By the same token, a fractal is a whole other breed of dimension.

First, it involves *effective* dimension. To understand this idea, imagine you're in Perth and planning a trip to New York. It won't do you much good to demand the truly shortest route to New York—because nobody's going to get a drilling machine and dig you a path through the center of the earth from Perth to New York. No, you're going to have to follow the curve of the earth. Although the solid globe has three dimensions, for long distance travel you can make effective use of only two of them.

Fortunately, you can get anywhere on the globe by using just two coordinates, latitude and longitude. You don't need depth, since there are no resorts or layovers at the earth's core. Even the explorers of *Journey to the Center of the Earth* got no further than a few hundred kilometers deep, although the core is about 6400 kilometers down. Thus we can forget about depth—it has no utility here.

In fact, when you fly from Perth to New York, you really need to consider only one effective dimension. How long is the string of your trip? Not how wide. Width doesn't matter here. The width of your body, the plane's width are irrelevant. You fly in only one effective dimension—length. Likewise, our Silly Centimeter vine grew in only one effective dimension, length.

In fractals, we count only the effective dimensions. The margin around the Koch snowflake has only one effective dimension. Like you on your trip, it too must zigzag back and forth. Its boundary line takes many sharp kinks along the way. But wait. Can a line, a length of one-dimensional string, have kinks in it? Where do they go?

Into another dimension. Well, partly. To a mathematician, they start into another dimension. But they don't get all the way there. They get only part way there, fractionally there. Snowflakes take these kinks in their boundaries in a regular, patterned fashion. These jogs repeating on different scales allow the process to be described fractally. The square snowflake has a bigger fractal dimension than the Koch snowflake because, what with being squared-off instead of triangled, it pushes its way farther into the next dimension.

Consider the fractal dimension of our snowflakes. The issue boils down to this: How much longer is the side of your snowflake when it jogs up to the next order of complexity? The side of a Koch snowflake jogs from 3/3 to 4/3 length. The Square Snowflake's side goes from 4/4 to 8/4 length.

We can make a ratio of this change. For either snowflake, cancel the common denominator of its old and new sides to form a new fraction using only its numerators. Then you read the fraction as logarithms. This gives us the fractional dimension or "shoulder-in tendency" of that change . . . its fractal, in other words. The fractal of the Square Snowflake is bigger than that of the Koch Snowflake.

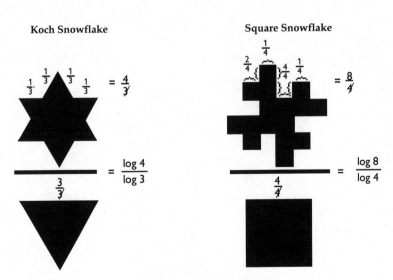

How long is the side of a snowflake?

The important thing is this: fractals make scaling ratios. You do it all the time, actually. If you say a penny is to a dime as a buck is to a tenner, you're using a scaling ratio. A Chinese might go fancy and say the 10 Chinese classics are to the 100 rays of lucky inspiration as the 1000 bounties of spring are to the 10,000 things spawned by the Tao. Both versions would use analog ratio to compare pairs of pairs, and both would use exponential growth.

Growth as Even Jumps

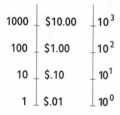

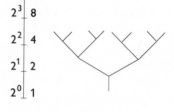

| 3 ways of saying the same thing by 10s | 3 ways of saying the same thing by 2s |

On the left of this graph, number grows in multiples of 10. It leaps from penny to dime to dollar to tenner in four proportional jumps up the ladder. It is exponential growth, going from 10^0 (or in other words, 1) to 10^1 to 10^2 to 10^3. But the graph to its right leaps upward in multiples of 2 . . . which abruptly recalls our old friend, the period doubling tree! Look at the vital difference here. Each progression here doubles the condition before. We may view it as period doubling, from 1 to 2 to 4 to 8 . . . *and* as exponential growth, from 2^0 to 2^1 to 2^2 to 2^3. Either approach will take us to the same number events.

In doubling, we jump up the graph in a way that also describes the wave frequencies in scaling octaves of music. Sound waves vibrate in exponential progression, so that if we treat a certain octave as our steady state (Period 1), then the next octave up gives waves that vibrate twice as fast (Period 2). The octave on beyond vibrates four times as fast (Period 4). And the next octave higher up (Period 8) has waves vibrating eight times as fast. Each new octave comes from an exponential growth of the original wave frequency.

Instead of graphing this progression upward in even jumps, we might choose to expand it proportionally to show the multiplying volume of growth. This will double the line length at each new period. Such a graph shows the exponential growth more clearly. It also makes us aware of how much richer is each new level, yet it is an organic outgrowth of the previous level.

Growth as Proportional Jumps

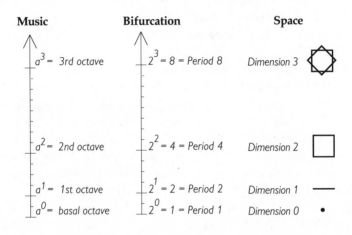

**Wave frequencies, bifurcation periods, & physical dimensions
all show exponential growth**

Scaling ratios are the heart of fractals. The spirals on the cover of this book have a twisty baroque look because real and imaginary number bend its fractal pattern elaborately. This creates a scaling that suggests an odd spatial depth. We'll see that the I Ching and the genetic code also show a strange sensation of depth . . . in time.

There is a central trick to all this bifurcation. The 2 is crucial. It is so versatile. It can go linear and add chunky units as 2 + 2. Or it can switch-hit into relationship—so here, as is typical for analogs, it offers a pair of choices, 2x2 or 2^2. But all choices bring us to the same result: our central, pivotal 4. It is the easiest of math.

But the 4 is even more crucial. The vital hinge between analog and linear realms is this pivotal 4. It is unique in the panoply of numbers because it offers a common goal for three different modes—adding, multiplying, exponential growth. That doesn't hold true for any other number. The dual state of 2 can take three different paths to this common goal, the magical 4. It jumps easily between linear and analog modes to weave a mesh of discrete linear units and analog proportions, so that "pairs of pairs" can compare and contrast.

Sheltered between the magic of 2 and 4 is the Period 3 window. It holds the hot spot. In a threesome, a pair can still compare and contrast in shifting relationship with the single. Like a human triangle stranded on an island, duos can form and reform, pitting two against one in various permutations. This is really why "Period 3 Implies Chaos."

A physicist recently told me that he'd always just taken it for granted that 2 and 4 have this unique relationship, just coincidence that plussing and doubling and squaring the 2 all bring the same answer of 4. But I think not; it's what makes patterned chaos go.

Chaos. Beyond the central 4-gate, the analog and linear roads diverge to different goals. Number must either travel the linear path of discrete units or the analog path of doubling/exponential growth. Answers will diverge further apart with each step.

Beyond the 4 we are forever split. The two forks debate over which choice is better. Even the analog path itself, being relational, has its own two forks—doubling or exponential growth—so that it forever frets between alternatives that are tragically burdened with irreconcilable differences. This is why, in the curt world of logic, an answer is either yes or no, but in the analog flux of relationships, as we are so sadly aware, reality is not nearly so cut-and-dried. In the holistic realm of the flowing psyche, there are tinges of ambivalence. It's both yes-*and*-no. There are two, even myriad sides to every story, process, issue, emotion.

Beyond the 4 we can't go home again. This simple fact has deep, unvoiced and often unconscious profundity. No wonder the Bible's Eden had four exits guarded by angels with flaming swords. No wonder the Taoists put four gates beyond the sacred center of their rituals. No wonder the four horsemen of the Apocalypse ride into chaos. Beyond the central 4, the linear and analog can never again mesh back into that Eden of sweetly coterminous number.

Oh paradise lost!

If analog and linear number no longer enjoy a simple union above the 4, then can they find a more sophisticated unity? They do. In analinear reunion. It is showcased in DNA and the ancient I Ching. Number pulls all into common purpose through a supersystem that counterposes one chaos dynamic against another. This simple, remarkable supersystem balances two Period 3 windows against each other—using number in both linear and analog modes—so that co-chaos is born. Stability plus evolution. In this paradigm, diversity moves back into unity.

Profoundly simple, this co-chaos system, so simple that it was first recognized by the Chinese perhaps 5,000 years ago. But as Paul Davis remarks in *The Cosmic Blueprint*, "The fact that the universe is full of complexities does not mean that the underlying laws are also complex." It is simple yet profound, this co-chaos supersystem. It codes for your flesh and your soul.

Soul? In science? A paradoxical new condition of our society is science studied with a mystic's heart. Science is discovering a new connectedness, an interrelatedness of events that suggests the universe is more than just random chance. Things in it mesh into a fuller, deeper meaning than the old Newtonian approach allowed us to recognize. There is scope for emotion here, for art, for meaning—for all those soft and vague "humanistic" qualities that science could not account for, but only count.

Physics, chemistry, biology—patterned chaos is drawing all the hard sciences into a new broadstream confluence with the softer subjects. Collaborations cross-fertilize each other, answers begin to appear that seemed unreachable just twenty years earlier. Chaos theory has developed its passionate advocates. Many scientists in various fields have turned their careers over to it, and for good reason. In fact some people show a nearly religious appreciation for this new window on reality, simply because it opens up everywhere and reveals its beauty so universally.

This holistic patterning touches the heart as well as the head. James Gleick says in his excellent general book on the subject, *Chaos: Making a New Science*, "The first chaos theorists, the scientists who set the discipline in motion, shared certain sensibilities. They had an eye for pattern, especially pattern that appeared on different scales at the same time. They had a taste for randomness and complexity, for jagged edges and sudden leaps. Believers in chaos—and they sometimes call themselves believers, or converts, or evangelists—speculated about determinism and free will, about evolution, about the nature of conscious intelligence."

An eye, an ear, a heart for pattern. A taste for complexity. The chaos converts. It suggests a new order of secular priests who honor a modern version of the Tao, where meaningful patterns trace out natural laws of quality as well as quantity. To recognize and honor both quality and quantity can put the whole world into balance.

All this sounds like the Taoist monks who studied nesting time cycles and found patterns on different scales in nature, who lived a philosophy of transcendent connectivity, who predicted wheeling patterns rather than specific events.

"What likes to go together?" the Taoists used to ask. Finding out will change your relationship with the pattern. And that changes everything. At each new zoom of increased consciousness the terrain alters; bridges form between one part of your life and another in ways you'd never expected; perspectives open on undreamed possibilities. Your life experience changes because you have changed your relationship to the pattern and thus altered the way that events synchronize around you.

Jung called synchronicity acausal because mere cause-and-effect logic can see no cause. But there *is* a cause—it's just not linear. Synchronicity works by analinear number. It taps into the spacetime flow and reveals patterns of holistic connection. Primal patterns—Jung called them archetypes—carry unique contents for each person. To see your own patterns and harmonize creatively and positively with them instead of misusing them will follow the way of the Tao.

In this invisible realm of the psyche, it is the proportion of things that matters. This determines the quality of your life experience, and it is why the ancient Taoists sought to tune into the cosmic design. They wanted to find the primal patterns in the flow of time and space. They strove to move within its amplitudes in such a way that harmony was furthered and disharmony minimized. They said that the attuned person waits for the right time . . . and acts.

The ancient I Ching expresses what Western physicists would call patterns of constructive and destructive interference. When two waves interfere constructively, they amplify each other's power to make a much larger wave. But when they interfere destructively, they diminish and may even cancel each other out. Here are several examples of interference in self-similar, repeating patterns. Notice that this basic pattern . . .

. . . can look seven different ways in the following strip, depending on how its overlaid waves go in or out of phase. Each pattern of interference emphasizes a different aspect of the form, much as seven personal relationships evoke different aspects of your psyche.

The first pattern is so destructive that you can hardly make out a design. But farther along come more constructive patterns that reinforce the basic design in pleasing ways . . . in my admittedly subjective opinion. Do you see a butterfly and a spider/bee in the second and third designs? I do. Analogs invite projection. They trigger networks of association, recall similar things carried in the memory. The last few examples to my mind actually improve on the basic design by making it more interesting. Thus, quality as well as mere quantity can come from the interaction of waves.

Interference patterns in the psyche will vary from person to person. Destructive interference can bring out an unappealing version of the archetypal form that invites a negative response in others. Or it can be constructively enhanced. The version that you manifest depends on what constructive/destructive interference comes from your own unique stance in life. What is Father? What is Mother? Your immediate association will tell you much about where you stand in the hologram. The way that you experience others—as Madonna or Whore or Madonna the Whore, as Wise Old Man or crotchety Senex or crusty Softie—all comes from your unique spot. It holds your unique interference patterns . . . because analogs always do. Your stance creates your own reality. You can modify it, though, by changing your mind. Literally.

How do you change your mind? Well, wave energy is expressed in anything flexible enough to wave: air, water, even the electrical jelly of your brain. These waves don't build in a linear way and add up their power by neat, even chunks of 1, 2, 3, 4. Instead the crests and troughs resonate in relationship so that one wave and one wave can make no waves or a so-so wave or a single huge wave—depending on their timing and spacing. Following is an example of this peculiar ability to increase or diminish wave power depending on its spacing and timing.

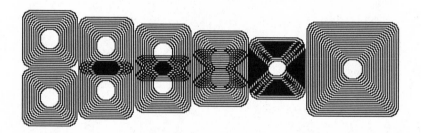

Here two waves merge to become one in pattern and purpose. This physical merger is a visual analogy for psychic integration. Converging currents of events need not tear your psyche apart, but instead can bring about a reinforced and larger unity.

Water in flux is the old symbol for the watercourse way of the Tao. Riding the chaotic wave of life to constructive purpose was the skill that the Taoists sought to learn and impart. In the laid-back jargon of Southern California, the Taoists were surfing the curl of emerging reality. The spiritual science of Taoism found life to be the ultimate wave, and living it well the ultimate ride. It taught that one can become a good surfer on the crests and troughs of chaotic events.

If life is the wave, then your psyche is the surfer. Stay one with the moving wave, thrilling and terrifying as that may be in the calamitous curl of events about to collapse over you in a wipe-out. Keep the right balance of straight-bore logic and circling analog habit that will allow you to shoot the tunnel of dark curving into light. You learn to ride the wave by studying your life's unique dynamic and following its swell and curve and break. This won't be easy. But it will be thrilling. And meaningful.

In Western society, we so often view life as quantity and less often as quality. We see matter physically but neglect it philosophically. We opt for the linear blinders of objective science. The gaze grows straight and sharp and focused . . . and misses so much. An example comes to mind from the 17th century. Dutchman Christian Huygens one day looked at a wall and saw that its array of clocks all had their pendulums swinging together in perfect synchrony. He knew that those pendulums had not, indeed, could not have been purposely set to such an exact synchronization. Huygens didn't develop a school of philosophy or monastic order because of his visceral thrill at finding time moving in such stately measured unity on the wall. Instead, he put his logical thought to work on it: how did those clocks entrain into a shared order, so that all the pendulums began swinging together? They must have transmitted their individual vibrations through the wall to each other.

Huygens saw it as a mechanical problem, but a Taoist would have seen it as a visible example of universal entrainment. Mechanical entrainment is only one level. Mental entrainment also exists. And there is yet another level where mind and matter merge into the universal flux within the order of the Tai Chi. More important still, one's manner of entrainment can change, and this comes through mere thoughtful attention. The ancient Chinese said, "I move my left hand and the whole world moves. You smile and it changes everything."

Since Huygens' day, Western science has discovered all manner of entrainments and mode lockings at the mezzo level around us, some of them quite bizarre. Women in a college dorm, for example, will tend over time to cycle their menstrual periods together. Order hides everywhere within the apparent chance of events. Not just order, but your personal meaning. Finding it can heal the split between objective and subjective, between discrete and holistic, between art and science, between matter and mind, between body and soul.

By merging the analog cycle with the linear unit, and motion with emotion, and head with heart, humanity can walk a transcendent third way. We are striding a broader and unifying highway into the future, and its name is patterned chaos.

WORKSHOP 5

Plan a program that combines discussion and experiences. Select book paragraphs to introduce the following topics—or your own.

Opening ritual.

❣ Consider your relationship history. In small groups, discuss what you attract and what you are attracted to. What specific traits have attracted you romantically . . . in friendships? These are your wells of fractal attractors. Why do you respond to these particular attractors? What attractors do you send out to others? Would you like to change any of this? How could you do it?

❣ The Bifurcation Chorus: count off 1 through 8 till all have at least one number. If the group is small, some may take double roles. The 1s go *flap-crack-snap*. 2s go *flap-crack-beat*. 3s go *flap-boom-bam*. 4s go *flap-boom-pop*. 5s go *buckle-twist-turn*. 6s go *buckle-twist-crinkle*. 7s go *buckle-shudder-shiver*. 8s go *buckle-shudder-shake*. Walk around the room making your own sound and movement. Find your closest relatives. Are there any with your own number? Number by number, sing and dance your part. Now do it in random sequence. You are a bifurcating tree busy at play. Now become part of a Period 3 window by joining two others numbers. Decide what topic your group wants to bifurcate—love, war, sports, whatever. Name your group and its forking actions. Do your routine for the whole group.

❣ Bring sheets of paper and drawing tools so that everyone may draw while sitting at tables or on the floor. Play music with a meditative quality. The task is to draw a yantra that will help you organize the major areas of your life into good synergy. Consider the shapes, colors, harmonies, flow of energy passing through it. Imagine yourself integrating all the parts into a successful holographic dynamic that makes your life function well. Draw this synergy into your very being. Show your yantra to the others.

❣ In small groups, bestow a public park on your city that would help people to experience harmony with nature. What would be the layout? What would it include? Who would go there? Why?

Closing ritual. Five minutes of feedback. Announcements.

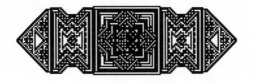

Section 2

East Meets West in Mind and Matter

It was King Wen that labored;
We according to his work, receive.
He spread his bounties;
Ours now to make secure
The destiny of this Chou.
Oh, his bounties!

Song 295 in *Shih Ching (Song-Word-Book)*
Early songs collected from 800-600 B.C.

Translated by Arthur Waley

Co-Chaos in the I Ching

Consider bifurcation as it develops in the I Ching. We begin with yin as ▪ ▪ and yang as ▬. In the West, thinking binary, we call it 0 and 1, nothing and something, no and yes. But in the relational approach of Chinese thought, yin is not just nothing. It is one of the two attractor points in Period 2 of bifurcation.

But yin and yang can move to a higher order. They rise into the four more complex states that the Chinese call the four bigrams: stable yin ▪ ▪, stable yang ▬, changing yin ▬, and changing yang ▬. Read them from the bottom up to see why they have acquired these names. (You always read the I Ching lines from the bottom up, since they grow that way.) A stable bigram remains the same going up to its next level, but a changing bigram alters into its opposite pole.

Thinking in Western style, a bigram is binary. It is just a couple of units stacked together. Stable yin ▬ is a stack of two black bricks. Changing yin ▭ is a stack where that second brick changes to white. In computer lingo, it switches from 0 to 1.

Westerners generally do look at the I Ching structure this way. In fact, many people see it as binary counting. Western binary counting was thought up by Gottfried Wilhelm Leibniz, the great German mathematician and philosopher. Leibniz sent news of his wonderful discovery to a friend in China, a Jesuit missionary named Joachim Bouvet. Bouvet must have smiled as he responded by sending back to Germany a copy of the I Ching.

What a shock! Leibniz was stunned. The binary code he thought he'd just invented had actually long preceded him in Chinese scholarship. It went back perhaps 5000 years in legendary Chinese history to Fu Hsi. Leibniz's own mathematical binary code, so modern, so linear and so logical, the very pinnacle of rational enlightenment—*this* was in the ancient I Ching?

Leibniz became greatly enamored of things Chinese and began to study the I Ching in earnest. In 1713 he eventually proclaimed it parallel to his own binary code in a book called *Two Letters on the Binary Number System and the Chinese Philosophy.*

Yes, Leibniz recognized the binary aspects in the I Ching, so let's consider what he saw. First of all, binary means *off-on*. The 0-1 state of alternation. Leibniz essentially saw yin and yang as binary code. When we consider things from his viewpoint, we can agree. This table shows yin and yang written in binary code. We are using yin ▪▪ as 0 and yang ▬ as 1.

Symbol		Binary		Decimal
▪▪	=	0	=	0
▬	=	1	=	1

And here are the bigrams written as strings of binary code.

Symbol		Binary		Decimal
▪▪ ▪▪	=	00	=	0
▬ ▪▪	=	01	=	1
▪▪ ▬	=	10	=	2
▬ ▬	=	11	=	3

Read a binary string like this: start from the right and move left. With 00, we have nothing in the 1's place and also nothing in the 2's place to its left—total count, 0. With 01, we have a marker in the 1's place, but nothing in the 2's place, for a total count of 1 unit. With 10, we have nothing in the 1s place, but a marker in the 2's place, for a total count of 2 units. And with 11, we have a marker in both the 1's place and the 2's place, for a total of 3 units. Either a marker adds in one more unit or it doesn't. This shows Leibniz's *either-or* approach, the chunky linear way of looking at the I Ching.

But we know there's also a relational, analog way to view this. So if you find math tedious, you can just scan this part for the analog "feel" of it. Consider yin and yang as two poles seeking a higher level of organization. In this approach, we are reverting to ancient Eastern philosophy—or forward to the modern West's chaos patterning. Here a bigram does not just stack one unit on top of another. No, that upper line of a bigram shows a higher order than the first line. It has reached a more complex structure, a higher level of organization. The upper line is in fractal relationship with the lower line. For instance, in the bigram of ⚎, the upper line is squared yin sitting in relationship with its lower yang line. Together they show a fractal rate of change.

All bigrams are fractals. Sometimes the second line will reiterate a polarity even more emphatically—but sometimes it will iterate the opposite pole instead.

Symbol	Doubling		Exponents
⚏	<u>doubly minus</u> minus	or	$\dfrac{yin^2}{yin}$
⚌	<u>doubly plus</u> minus	or	$\dfrac{yang^2}{yin}$
⚎	<u>doubly minus</u> plus	or	$\dfrac{yin^2}{yang}$
⚌	<u>doubly plus</u> plus	or	$\dfrac{yang^2}{yang}$

The result is that the second line doubles the first line's intensity. To understand this, recall our square snowflake. It reached a second level of complexity when we made it "vibrate twice as much" by jogging it in and out in a regular way so that it became 8/4 long compared to the old 4/4 version. It doubled the line's frequency without extending it at either end. It shows a fractal rate of change.

And the same occurs with the bigram! Its upper line is twice as complexly organized as its lower line, vibrating an octave higher than the bottom line. It is a ready-made ratio, and already it uses position to show that its upper line is at a higher level compared to its lower. Both lines belong to the same process; they are just two different stages in it. And this complex polarity is analog. Its poles do not even exist except through relationship.

By this evolution into the bigrams, now we're no longer stuck at that earlier crude level of simple yin and yang. To make an analogy, we've moved past black and white to include two shades of gray. We've developed four attractor points for the eye.

Psychologically speaking, reaching this new level means that we are beginning to realize there's some good in the bad and some bad in the good. Things no longer look so black versus white, villain versus hero, yes versus no, innocent versus evil. Things evolve and complicate. It is why the ancient Tai Chi symbol put that continual reminder of the yin dot birthing within the yang and vice versa.

Now let's move up to the next level of bifurcation. Here our four bigrams become further nuanced into the eight trigrams. They too can be seen as binary or analog. The black and white stripes in the center show an easy way to make binary trigrams. When you run vertical splits down through the top row of stripes, you get eight stacks of unitized black and white bricks . . . easy to code into yin and yang. On the left is a stack of equal black bricks that is all-yin.

Two Ways to View Trigrams

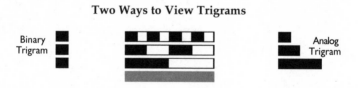

But the analog mind doesn't see these levels as equal chunks. They are octaves. In the analog view, this trigram's bottom line has the deep resonance of a long string vibrating its slow base note. The middle line is an octave higher, vibrating twice as fast. The top line's yet an octave higher, a treble frequency vibrating twice as fast again. Each is a different wave frequency in fractal ratio with the others.

The outcome is that we have here two distinct yet complementary ways to consider a trigram. Analog mode and binary mode. Yet paradoxically, both modes come from the same bifurcation tree!

Consider the two views on the next page. The linear mind reads *across* the tree, counting its number of branches in a horizontal slice. It counts the chunky units of tree cordwood. This gives the *horizontal* Period 3 window that is common to modern chaos theory. When only three branches appear in a horizontal window, each branch with its 0 or 1 binary value, you get a trigram. The numbers here represent a specific trigram. For example, 001 is the trigram of Mountain ☶. Thus you can have only one Period 3 window at that level, since it has only three branches. But of course, since there could be eight possible windows, there are eight possible trigrams. This binary method reads power across the tree.

But viewing the tree in analog mode follows one branch upward through three levels to create a *vertical* Period 3 window. This view gives us eight different windows at the top, each a trigram. Each extends downward through three levels to create a specific fractal trigram. It reads power up and down the tree.

Most important, either approach shows why a trigram only needs three lines. It contains the whole key to patterned chaos, the Period 3 window!

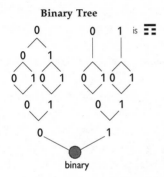

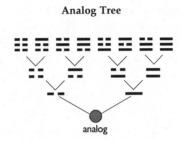

A single trigram (001 or ☶) as a horizontal Period 3 window. It reads across the tree top.

The 8 trigrams seen as 8 vertical Period 3 windows. It reads up the tree forks.

Through either approach, we reach the colorful family of trigram relationships. Color is a development rising beyond black and white, or even the two added grays of the bigrams. Eight trigrams lie along the color spectrum, from all-black yin ☷ of Mother Earth on through the range to all-white yang ☰ of Father Heaven at the other end.

Notice that the masculine and feminine principles are different but equal, engaging in complementary creation. Creative power is not just a heavenly "he," nor is it a global Gaia-package of "she." Father, represented by the encircling sky, is no more powerful than Mother, represented by the firm, foursquare landscape. The trigrams of Heaven and Earth generate six children—or further trigrams—affecting the environment through their various analinear powers.

This Old Family order you see below counts in binary code from 0 to 7. It's the same order you saw in the stripes that split into bricks. Some years ago I assigned colors to these trigrams in keeping with the paint spectrum and the four elements of Earth ☷, Water ☵, Fire ☲, and Air ☴. This color order must exist within the collective psyche, because I found it has occurred to others too. Eventually I even saw a deck of little I Ching cards on the market using this same color format.

This Old Family order of trigrams comes from Fu Hsi inventing the Ho Tu, that ancient map he discerned in a spotted pattern on the back of a "dragon-horse" rising from the Ho River. You'll find more about this map later, in the chapter called "The Atomic Map."

The stable Old Family makes a lovely pattern of relationships. Each trigram has its own name and traits. It is unique yet complementary. Its nature can be intuited from meditating on its name, image, polarity setup, and color. But some of this is quite subtle.

Consider the oldest daughter. She is named Wood-Wind ☴. The Western mind might find it hard to see any link between wood and wind, but the Chinese mind thought they shared a remarkable trait. Wood-Wind has the tendency to introvert and hide invisibly behind its effect. This shy yet competent oldest daughter creates lasting change through the gentle persistence of her self-effacing strength.

One cannot see the wind, only what it moves—leaves, dirt, smoke, banners, rain, snow. But the wind can carry tons of sand or snow in its invisible grasp, can etch canyon bridges and build glaciers. By the same token, the casual eye does not notice a tree growing larger from day to day. But cell by cell, its woody volume slowly swells from a tiny seed into a stem, until a tree even as huge as the sequoia emerges from this quiet, long-lasting effort. Both wood and wind suggest the gentle, inconspicuous persistence of this oldest daughter, ☴.

Old Family

Black Earth	Purple Mountain	Blue Water	Green Wood-Wind	Yellow Thunder	Orange Fire	Red Lake	White Heaven
☷	☶	☵	☴	☳	☲	☱	☰
Mother	3rd Daughter	2nd Daughter	1st Daughter	1st Son	2nd Son	3rd Son	Father

There is an odd philosophic quality to studying the trigrams. Perhaps you notice it from this example of Wood-Wind. All depends on analogy. Number has analog quality here; it instills relational wisdom rather than emphasizing linear knowledge. The I Ching must be learned slowly and it brings a quiet heart enriched by meditation, rather than the glittering bravura and klieg lights found at the linear centerstage of attention. So be patient with these trigrams, get to know them slowly, treat them with deep relationship as well as logical thought, and they will spiral you into a new awareness. They will reward you more profoundly than any set of linear facts alone.

Let's consider these family members sitting on the color wheel, with the parents in the center and the siblings sequenced around the paint spectrum in an orderly ranking by birth. Across the wheel, pairs of trigrams act as mirrors in complementary ways.

First, these pairs reverse each other by color. If you stare fixedly at purple, for instance, you will get an afterimage of yellow that is the complementary partner across the wheel. This color pairing is consistent across the Old Family wheel. Next, these pairs are also polar complements. For instance, ☵ and ☲ reverse each other in polarity, as do the poles of all the other cross-pairs. Finally, the pairs are also reversing mirrors of poetic imagery. The mountain rises up while the lake hollows down, fire burns while water extinguishes, a sharp crack of brief thunder brings shock but gentle wind soothes in the enduring forest. Bright heaven lofts clear with the sun's rays, while dark earth cycles opaque matter in pregnant mystery.

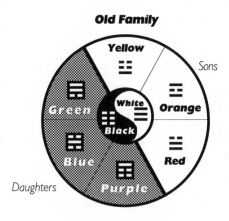

Old Family

This strong mesh of relationships in the Old Family is stabilizing and orderly. Its patterns have infinite philosophical depth, recalling the fractal windows of the Mandelbrot set. The closer you look, the more you will find. Hellmut Wilhelm remarks in *Heaven, Earth, and Man in the Book of Changes,* "The system of The Book of Changes is the representation of a multidimensional world. Pairs of opposites should not be looked for only at the poles of a one-dimensional axis. Depending on the direction of view, there will be found a number of different opposites for every given concept or situation."

These concepts can even be put to scientific use. The Old Family order counts in binary code from 0 to 7. To put it into a Western-style format, we'll use ▪ ▪ as 0 and ▬ as 1. Frankly, to do it, I mentally tip a trigram over on its side with the top line to the right—for instance by comparing 000 to ⦙⦙⦙, or 001 to ⦙⦙⦚. Or you could instead mentally flip the number string sideways. Both techniques show the parallel.

Trigram		Binary		Decimal
☷	=	000	=	0
☳	=	001	=	1
☵	=	010	=	2
☶	=	011	=	3
☱	=	100	=	4
☲	=	101	=	5
☴	=	110	=	6
☰	=	111	=	7

Binary, to be sure. But, science can use the trigrams in an analog way, too. Each rising line doubles its polar complexity. Below in the trigram of Fire, Line 1 is yang; Line 2 is doubly yin; Line 3 is yang doubled yet again. Moreover, it can also be viewed as exponential growth! Each rising line is a higher octave that rings or counters more intensely the line just below it. Thus the second line becomes squared yin, while the top line is cubed yang. This vertical Period 3 window is much more interesting than its horizontal version. It is in effect a triple-decker fraction, a "strange attractor" fractal.

$$\text{Fractal Trigram} \;=\; \equiv \;=\; \dfrac{\text{yang}^3}{\dfrac{\text{yin}^2}{\text{yang}}}$$

Each trigram describes a program holding its own kernel of chaos patterning. Each trigram describes a different set of fractal attractors that can be used to discuss a configuration of spacetime and mattergy. It can even discuss the flow of your psyche. Thus, this Old Family order provides a basic structure of polar arrangement. It gives a stable foundation. But these fractals known as the Old Family tell only half the story, just half of the paradigm of co-chaos.

Co-chaos also uses the *New* Family. Yes, there's another family. Actually, both families contain the very same trigrams, but they sit in different sequences of polarizing rhythms. It is as if two families live near each other, both with two parents and six children—such big families in those old days! Similarities do exist between these families, yes, but there are also differences.

The New Family is more volatile and changeable. It is less steady but more innovative. It is more on the go, but with less visible means of support. It marches to the beat of a jazzier drummer.

What a strange order this is! It does not count out the binary code from 000 to 111. Mother is still on the left and systematically changes into Father Heaven at the other end. But it happens through a whole different rhythm! And how strange! Here the second and third siblings have exchanged gender! Now Fire and Lake are daughters, but Water and Mountain are sons. The opposite pairs are still in color complement. But on a different spectrum—the light spectrum!

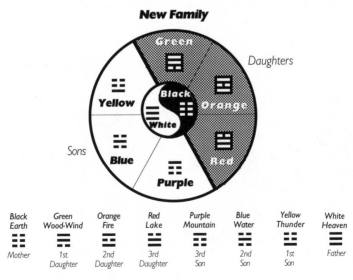

Black Earth	Green Wood-Wind	Orange Fire	Red Lake	Purple Mountain	Blue Water	Yellow Thunder	White Heaven
Mother	1st Daughter	2nd Daughter	3rd Daughter	3rd Son	2nd Son	1st Son	Father

Interestingly enough, the Old and New Family orders correlate with the paint and the light spectrum. How so? Well, tilt a cube beside each color wheel and transfer its primary colors onto the cube's projecting corner and those three edges. The Old Family cube shows the paint spectrum merging into black, but the New Family cube shows the light spectrum merging into white!

This congruence of color systems suggests new possibilities for explaining some of the puzzles that remain in the not-as-yet understood process of human eyesight. Complementarity has long been noticed in eyesight, for instance, in Hering's "opponent-process theory" of the 19th century. Optometrists and opthamologists work in two different systems of vision correction. And it is not yet known exactly how the three kinds of vision cones connect to the bipolar cells, but some sort of complementary polarization is indicated.

What odd sort of complementarity is embedded in the Old and New Families? It is based on number relationships within the two ancient maps of the Ho Tu and Lo Shu. While the Old Family provides cycling stability, the New Family introduces a syncopating counterbeat that energizes and alters that steady rhythm. Together, they develop a gloriously complex yet fundamental heartbeat which on the linear side can count in 2- and 5-base, but on the analog side uses period doubling and exponential growth. Such a solid mesh of number functions provides the fail-safe system that insures the infinite backup necessary for the cosmic dynamic of co-chaos. Here simplicity generates complexity . . . and vice versa.

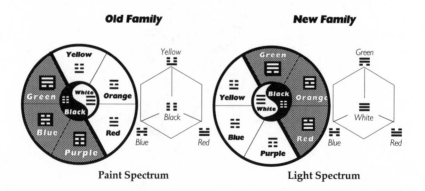

Old Family   **New Family**

Paint Spectrum   Light Spectrum

What practical use are these analinear trigrams? One possibility is in geometry. I believe this complex polarity shows the reason why we have only three dimensions of space, and also why we live out time as its complementary trio of past, present, future.

Consider. A trigram can describe a position in 3D space . . . or time! For instance, the polarity of space progresses from point to line to plane to cube, or from gray dot to yin/yang to bigram to trigram. In the diagram below, notice the layout of poles at each rising level. This polar sequence isn't arbitrary or random. Modern math and the ancient Old Family compass mandala have convinced me that each pole must reside in just this location to work right. A simple algorithm puts each pole into its own trigram. For the cube, the rule is *width* pole = bottom line; *height* pole = middle line; *depth* pole = top line. And the Old Family ordering around the cube comes from the two Z movements in *depth* given by a doubled 4-beat that is shown in the Ho Tu dots, joined by its fifth-beat-pause in the center.

Polarized cubes can join in lattices to make 3D cellular automata sending complex messages throughout a system. They can describe frequencies ringing at the subatomic level to resonate constructive and destructive interferences. And by configuring a spacetime cube with a mattergy cube to give the hexagrams, you get a cosmic hypercube. I suspect that hidden here is the core of string theory, our latest attempt in theoretical physics to follow the music that beats out the rhythmic changes of matter and energy in space and time. This surety of interlocking number meshes into an analinear pairing of vertical/horizontal Period 3 windows. By combining trigrams into hexagrams using these hypercube phase relationships, the I Ching runs an eternally dynamic program in the complex polarities of bifurcation. It is cosmic mind discussing cosmic matter.

Building Space With Polarity

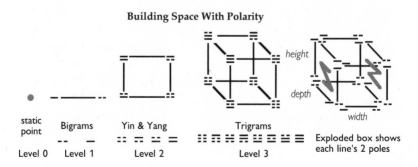

125

We have seen trigrams as binary and analog number, as fractals and philosophy. But just how do these trigrams fit together to make complementary chaos, or co-chaos for short? The pairing of the eight trigrams is what the I Ching rises to in its next level of development: the 64 hexagrams. You can see them on the opposite page.

A hexagram is six lines deep. English *hexagram* comes from Greek *hex* meaning six and *gram* meaning line. Actually, technically speaking, even the Star of David is a hexagram because it has six lines. But just as the Star of David is really two interactive triangles, not just six lines, likewise, the I Ching hexagram is two interactive trigrams, not just six lines. In a hexagram, these trigrams represent two Period 3 windows that show two chaos systems counterposed in dynamic balance. Working together, they form a remarkable supersystem of complementary chaos, or co-chaos for short.

In co-chaos, the Old and New Families pair their trigrams through all the possible permutations to make the 64 hexagrams. The math-minded will notice that this means the basic plan is not just 2^6 lines = 64 hexagrams, but rather, it is *also* $2^3 \times 2^3$ lines = 64 different patterns of finely tuned co-chaos. The analog view of the hexagram demonstrates exponential growth's best gift, the paradoxical fact that you can add exponents to do multiplication.

In a hexagram, each trigram holds its own internal recipe for chaos. Its pivot is the central line that allows the three different modes (adding, doubling, and squaring) to use the goal of 4 as the home base. No wonder the ancients taught that a trigram's most important line is its central line. These central lines provide the fulcrum of number synergy that allows the complicated rhythm of linear and analog functions to create the co-chaos supersystem.

The opposite page shows the 64 co-chaos equations all together. Of course there are many possible ways to lay out these hexagrams, but this particular order has a special significance. I call it the stable Old Family order, but it is also known by other elegant designations such as the Primal or Early Heaven or Pre-Heaven order. So many alternative names for the same thing can be confusing, I admit. The ancient Chinese loved nature analogies, as you can tell by their poetic naming, and an analog mindset always likes to hold alternative choices. All becomes allusion, metaphor, holistic analogy. I have chosen this Old Family name, though, to emphasize the way it uses the stable order of Old Family trigrams.

Notice the vast and profound regularity here. In each row going across the chart, all the upper trigrams are in Old Family order. Likewise for the lower trigrams running down the chart. It all shows a simple and universal symmetry. Separately these upper and lower trigrams count out their binary rhythms; then, they in turn feed into the larger binary pattern of the 64 hexagrams. They count out the whole chart from the binary 0 of Hexagram ☷ in the upper left corner to the binary 63 of Hexagram ☰ found at the lower right-hand corner. (Note that we're counting from 0 to 63 here, not from 1 to 64.)

Old Family Hexagrams - Binary Code of Leibniz

Hexagram Title	I Ching		Binary		Decimal
The Receptive	䷁	=	000000	=	0
Breaking Away	䷖	=	000001	=	1
Holding Together	䷇	=	000010	=	2
Contemplation	䷓	=	000011	=	3
Enthusiasm	䷏	=	000100	=	4
Easy Progress	䷢	=	000101	=	5
Gathering Together	䷬	=	000110	=	6
Standstill	䷋	=	000111	=	7

The Old Family hexagrams of course show the same binary order that Leibniz recognized. To him these hexagrams looked like six broken or unbroken lines stacked into various sequences of *off-on*, counting out their linear, discrete, additive chunks of information. He apparently didn't see the structure as analog at all. Nevertheless, he was greatly influenced by the philosophy and implicit harmony of the I Ching . . . to the point that Voltaire tried to make Leibniz look like a fool by satirizing him with witty brutality in *Candide* as Dr. Pangloss, the blithering professor who is forever proclaiming amidst dire predicaments and great sorrows that "All things are for the best in this best of all possible worlds." And finally, Candide finds the Tao in his own garden. Of course, to a brilliant and facile-tongued cynic like Voltaire, the dark and wordless way of the Tao might have seemed senseless indeed. True, it's not merely logical, but it is whole.

Each of the 64 hexagrams is a unique meshing of linear and analog functions. Each shows its own distinct dynamic pattern in a short-form equation (in terms of calculus), or as a recursive algorithm busy iterating its self-similar patterns on scales large and small (in terms of dynamic chaos and cellular automata). Each hexagram puts two systems of strange attractors into relationship and sets up an extraordinary rhythm that uses both linear counting toward product and analog recursion of process. Not only do the I Ching hexagrams describe the 64 basic states of co-chaos, but they also even model its ability to change from one condition into another by means of this cosmic heartbeat based on number.

We notice this beat of the archetypal bifurcation numbers—the 2s, 4s, 8s and the paired triplets of co-chaos—echoing on various levels of reality. For instance, we glimpse it in the versatile pivot of the 4-valence that allows the carbon atom to become the basis for our planet's life. The periodic table of atomic elements uses the octaval range of Period 8 bifurcation to order itself; in the atoms, this rhythm is visible in their electron shells. And what of the eightfold way of the subatomic particles? What of the paired trios of quarks that echo this same theme of paired triplets that we have found in co-chaos?

I suspect that this deep co-chaos dynamic is the loom of reality. Its eternal beat weaves the cosmic fabric. It is what gives matter the energy to group and regroup itself in space through time to generate the universal changes of both evolution and entropy.

Exploring these notions has led me to apply co-chaos to that most fundamental pair of pairs, the complementary duo of spacetime and mattergy. A real Theory of Everything—or a TOE—it seems to me, would include not only the physical realm with its gravitation, the Planck constant, and so on, but also that ineluctable realm of mind with its intangible qualitative values. All this intrigues me into suggesting some possibilities that this book, focused as it is on organic life, cannot pursue very far here. So exploring that is left for another book called *god's TOE*.

Sure, Leibniz saw his binary code in the I Ching. It was there. But there is so much more. As we have seen, it also codes for analog relationship. Number expresses both its linear and its analog modes here, meshing into the paradigm of co-chaos. All this brings an analinear balance to number which makes it exceedingly primal and powerful. This archetypal number, with its beautiful and useful analinear weave, is strong enough to hold the shape and weight of the universe . . . and of universal meaning.

WORKSHOP 6

Plan a program that combines discussion and experiences. Select book paragraphs to introduce the following topics—or your own.

Opening ritual.

❣ Bring several I Ching books to the session (more is better). Give out sheets of paper and ask people to write down their ages. If you're 63 or under, the I Ching has a hexagram numbered with your age—for instance, age 38 has Hexagram 38, *Polarity,* ☲ . . . or if you're older, pick a relative's or friend's age. Pass around the I Ching books and let everybody find their age hexagram and study its basic description. In pairs, discuss your hexagrams. Through association, what might it suggest about you? Does any of it really apply currently?

❣ Ask people to write down their ages in binary code. If necessary, they can help each other. What if we had to write out all of our daily numbers in this way? Discuss the pros and cons.

❣ What is your favorite color? Which trigram would it be in the paint spectrum? The light spectrum? In pairs, discuss your favorite color and its two trigrams. Combine these two trigrams into their two possible hexagrams. Which ones are they? Look them up in the I Ching book. Do they suggest anything to you about your basic nature? Your potential behaviors? Use these mental associations as a subliminal cue to help you open up to some new perspectives on your habitual stances in life.

❣ Meditation: Consider your age. What tasks are you facing this year? What dynamic do you want to incorporate this year to deal with your tasks well? Does that seem to be more yin or yang? Ask the Tao to help you recognize and develop the dynamic pattern that you will need to succeed in this year's tasks.

❣ Stand in two facing lines. Think of an occasion when it helps to access your linear logical mode. In orderly sequence, let each member in the row name an occasion in a phrase to the opposite side. Next, stand in a circle. At random, conjure up and call out moments when it helps to tap into your analog mode. Design a short ceremony to affirm and integrate both modes.

Closing ritual. Five minutes of feedback. Announcements.

Co-Chaos in the Mind

The I Ching works as subtly as the mind itself. Like a microfilm viewer that can range from broad-scan to fine-focus, its answer can encompass thirty years of your past or five minutes of the future. With it, you move by resonances among the nested cycles of the Tao. It is the analinear dynamic that allows it such great scope of resolution, but only your experience and well-kept records can show you whether it really works. Each answer is an analogy for you to apply to events in a subjective way. Its accuracy is not a matter for blind belief, but instead, for your empirical experience to assess.

The I Ching speaks at the borders of consciousness, so it's quite easy for the sharp-edged ego to dismiss an oddly apt answer as some weird coincidence due to a momentary suggestibility or mental quirk. Even worse, a fearful or self-protective ego can label a puzzling answer as just pointless garble. You must *observe* the congruence happening over enough time, through your own records kept and re-examined with some distance from the heat of an event, to know it works. You must seek truth enough to want to drop the projections and be honest with yourself, before ego can even relax and let truth in. . . at least if you're the skeptic that I used to be.

How can some overmind look both backward and forward with the focus on your individual psyche? Well, it's no more mysterious than the way that the genetic code can look both backward and forward from the focus of your individual body.

In consulting the I Ching for an overview on the past, present, or future, you in effect become a time traveler. Sounds absurd? Well, scientists already know that the genetic code travels in time. It travels the entire span of the human race, and indeed of all life on earth. It deals with the past as a time capsule holding the summary of our whole heritage starting with the first living cell. It preserves in itself the evolving history of each species. Though the individual members die continually, the genes carry on.

Genes also plan for the future. For each species, the genetic plan keeps adding to itself, developing its potential to build a better, more diversified creature or plant or virus. The history coded inside the genes keeps diversifying and lengthening since the living system is in a constant state of adapting. Organic life constantly changes while still maintaining its momentum.

This ability of life to reproduce and complicate itself endlessly is amusingly discussed by Douglas Hofstadter in *Metamagical Themas: Questing for the Essence of Mind and Pattern*. He says, "Ever since self-replicating molecules came about, they have been reproducing like mad and proliferating in ever more varieties . . . . We ourselves are huge self-replicating molecule-heaps. Ever upward builds this dizzying spire of self-replicating structures. What gives this whole movement any coherent direction? How and why does complexity evolve from simplicity?" I would suggest that the answer may be found in co-chaos. Its simplicity allows nature's complexity.

The organic momentum toward more specialized structure has long puzzled scientists because it appears to contradict the second law of thermodynamics. This law, a product of mechanistic 19th century thinking, states that in any process, some energy dissipates as heat. We can see this law in action in fueling a car. No fuel can be used with 100% efficiency to generate motion. Unwanted heat is also generated along with the motion.

The second law of thermodynamics says that the entropy of an *isolated* physical system must keep increasing—in other words, its disorder will increase as its energy decreases. If the living system were truly isolated and merely physical—not connected to other systems and with no spiritual component— it should deteriorate and fall apart by losing its energy through its various processes.

But we see that the opposite is happening: while individual units do indeed deteriorate and fall apart—people, horses, frogs, bacteria, whatever—the living system keeps on keeping on. It continues evolving in its many species, branches, and sub-branches that crowd each other for new windows of opportunity in a complex ecology. For example, more people mean fewer elephants; fewer people mean more insects.

We can readily see that life becomes more diversified as it also maintains momentum. This trait is called *epigenesis* (from Greek roots meaning *after* and *creation*). It means that life keeps on creating itself and elaborating on itself in ways that you couldn't have predicted from looking at the original cell at its start. Erwin Schrödinger commented that the living organism seems to feed on its own increasing complexity. This trait is called "negative entropy"—or neg-entropy for short.

By feeding on the energy of stars, bacteria, money, gasoline, and eggs Florentine, our human life proliferates and increases in complexity. But we are only the latest step in a whole cosmic trend that has already taken many earlier steps up to a higher order. Cosmic matter "froze out" to various symmetries of increasing organization, from hot quantum soup to quark-photon potage to hadron hash to nuclear nuggets to atomic bonbons with their hazy electronic coating to the molecular crazy salad of compounds that eventually led to living organisms. All this increasing order took time.

The genetic code is a knowledgeable time traveler. It has existed since the beginning of life, it compiles and sorts data in the present, and it makes decisions for the future direction it will take. Biologists have now discovered that life's evolution rate is far faster than would happen if species change occurred merely through random mutation. There seems to be some sort of foresightful or "purposeful" mutation that goes on. Thus the gene carries within itself the past, present, and future of all living things. Containing all the past in capsule, it packs ready to jet into the future.

One can explain this odd impetus of evolving life by simply stating that the living system is not isolated. It has connections to everything else. Lila Gatlin mentions in *Information Theory and the Living System* that classical notions of entropy are simply inadequate to explain the complexity of the living system. She notes that it is useful to bring the tools of mathematics to bear on this problem. But she also wryly remarks, "Mathematics is simply a way of expressing concepts that anyone can understand in a way that very few can understand." So let's not get fancy with math here in a way that very few can understand, when it isn't even necessary. We can stick to the simple numbers like 2, 4, 6 and 8. We'll use the I Ching itself as a powerful mathematical tool combining linear and analog modes.

The genetic code is actually one manifestation of a much greater mathematical principle in operation. Gatlin foresaw in 1972 that, "The genetic code is a small subroutine of a master program which directs the machinery of life. We have no idea what the language of this master program is like, but we can be sure that it has always evolved, is now evolving, and will continue to evolve in the future."

Co-chaos suggests that the whole cosmos is alive. It evolves not just in what we call life, but at all levels. There is the inorganic—as nuclei turn into elements into suns into solar systems into galaxies into superclusters—and organic—as microbes proliferate and whales school and ants nest and people lounge about in Roman baths or health club saunas. Entropy and evolution are partners. They balance in the endless algorithm of the universe.

Some people call this master program a universal mind or a cosmic consciousness. How to describe the divine, the oversoul, the ultimate one? Heisenberg notes in *Scientific and Religious Truths* that the often-quoted sentence of "God is a mathematician" really derives from Plato. He says that modern mathematicians have chosen to confine their vision to the realm of mathematical proofs, but Plato himself did not: "Having pointed out with the utmost clarity the possibilities and limitations of precise language, he switched to the language of poetry, which evokes in the hearer images conveying understanding of an altogether different kind . . . the language of images and likenesses is probably the only way of approaching the 'one' from more general domains."

In the West, the all-encompassing pattern of god has usually been left to the artistic, the creative, the mystical contingent to explore, considered by general society to be unprovable and unscientific. But still, even scientists have had their occasional flirtations with the numinous mystery of huge pattern beyond liner logic. Newton, for instance, was an accomplished astrologer. When someone attacked his reliance on astrology as a foolish superstition, Newton replied, "I, Sir, have studied it. You have not." Leibniz wanted to redefine the I Ching in Christian terms. He sought to formulate a philosophy that would also scientifically prove the existence of god.

Hans Driesch in the 19th century developed a "vitalist" theory which postulated that some non-physical cause must give life its vital force. He called it *entelechy* and said it holds a goal toward which the living system is directed. He said it's a non-spatial cause which is, however, enacted into space, and he pointed out that this unknown form of power doesn't contradict the second law of thermodynamics. Driesch even hypothesized that it works by affecting the detailed timing of microphysical processes, somehow suspending a process and then releasing it from suspension when the process is "set free into actuality . . . ."

Unfortunately, in Driesch's day, the mind-bending quirkiness of quantum mechanics had not yet been discovered, and entrenched Newtonian physicists, with their emphasis on hard-headed reality, dismissed his vitalist idea as unsubstantiated wooly-mindedness.

Rupert Sheldrake, however, in *A New Science of Life* points out that by modern standards, Driesch's vitalist theory is not so full of holes now that the unpredictable science of quantum physics has arrived on the scene. "This theory is by no means vacuous, and could probably be tested experimentally . . . ," he says. Sheldrake suggests that some set of scientific variables may explain the nature of entelechy and how it works, but we just don't know it yet. He postulates some kind of gene-changing energy that he calls morphogenetic fields. But until we do know how it works, he says, any vitalist theory will be unsatisfactory because entelechy necessarily operates on the physical world from a realm of networking connections that we do not yet understand.

James Lovelock furthered this vitalist position by developing a thoroughgoing presentation of biological evidence for the earth itself as the giant living organism he calls Gaia. His books describe our globe as a huge organism which lives, evolves, and responds to outer stimuli and inner process in a coherent and self-regulating manner. In a manner of speaking, we are bacteria in the belly of Gaia. And the scientific community at large is beginning to think that perhaps this may really be so.

But the vitalist theory, morphogenetic fields, and Gaia are just new ways to express an old, old idea. It has existed in many cultures, this view of the earth itself as a living organism. In Thomas Berry's words, "Although this belief was never central to Western thought tradition, it maintained itself consistently on the borders of Western consciousness as the *anima mundi* concept, the soul of the world." Along with living in the bosom of the world's physical body, analog cultures have felt cherished by a world soul, a tissue of invisible connection. Yet this connective, holistic mindset became disparaged in Western culture ever since the Greek heyday of linear logic began to win out over the mystery cults.

Is the I Ching a mystery cult? Why not find out for yourself? Try it for three weeks. Really ponder your questions . . . and answers. Look for a connective resonance beyond the causal factors. Keep records and check back to verify. If your logical, linear left brain does not overwhelm and shame your holistic knowing, I suspect that time and experience will show you a connective truth far beyond chance. You may even find a wise friend.

What happens when you consult the I Ching? A complex beat of fractal attractors in spacetime focuses on your mattergy, tuning right down into your own mindset to give you a specific hexagram that resonates with your issue and shows you its dynamic pattern. This answer is to be carried and turned about in the privacy of your own evanescent thoughts. The I Ching is just that subtle. The cosmic network of reality is just that thoroughgoing and complete and tuned to all. If you suppose that god's eye is on the sparrow, it should be no shock to discover that god's mind is on yours.

Your psyche, your body, your culture, your species change. The universe itself lives by generating change. Change is what the hexagrams describe: change in universal mind, change in universal matter. It is all a giant living organism. And you are part of it, capable of finding your path, capable of tuning in on the mind of god as it relates to you.

Since the living Earth is sometimes called by the name of Gaia, it amuses me to call the living universe by the name of Cosmo, which in Greek actually means universe. Cosmo is universal mind and matter, alive and well and working at every level of reality—the tiny fraction that we perceive and all the rest that we don't. It operates in an immeasurably huge and complex system. It is based on the co-chaos paradigm whereby the primal pair of pairs—spacetime and mattergy—move through the infinite permutations of co-chaos in a huge flux of spiraling change.

The ancient Chinese coded the co-chaos dynamic into the I Ching. But how? How was it developed? Three great rulers usually get the credit—Fu Hsi, Yu, and Wen. But Chinese history is full of rulers who were given mythic accomplishments that often actually sprang from many smaller discoveries made in the collective. King Wen was the man who wrote down the I Ching we now have, and he is the hero lauded in the folk song on page 113 prefacing this section called *East Meets West in Mind and Matter*. King Wen has such a fabulous history that I want to tell you some small part of it.

This man became feudal lord of the Chou tribe in China back in 1185 B.C. He headed up this rustic and rural tribe in the west, but nevertheless he enjoyed a rather good relationship with his more sophisticated overlord to the east, the Shang king. You can read "sophisticated" here to mean corrupt and degenerate to the simpler eyes of the Chou people. And indeed, the Shang rulers ritually slaughtered hundreds as sacrifices that were buried in the royal tombs. They drank to stupor, tortured for amusement, and spent huge amounts of tax money.

Nevertheless, our rustic hero from the west rose to some power, and he even had the honor of receiving a bride from the Shang royalty. You can find the mention of his wedding and the Shang king in the I Ching, specifically in line 5 of Hexagram 11 and line 5 of Hexagram 54—unless your version edits it out under the premise of "who cares about old names?" In English that corrupt ruler's name is variously spelled as Diyi or Ti-Yi, or leaving off his title, sometimes just as Yi or I. Many are the vagaries of taking characters from a picture writing into the very different format of alphabet writing. Generally I pick and choose among the possibilities in the backlog of historical precedent. For instance there is surely only one way for our English language to spell Confucius, no matter how the latter-day fashions may veer format—for instance, currently the People's Republic of China prefers "Kongzi" in Pinyin transcription.

Anyhow, the tyrant who followed Yi in rulership, the last king in this dynasty, was notorious for a wanton sadism that was extraordinary even in the Shang empire. For instance, one winter day this cruel King Shau (or Cheao) saw some peasants wading through a cold stream. He ordered their legs cut off at the shank so that he could "inspect the bone marrow that could endure such cold." And once when a relative rebuked him sternly for his cruelty, King Shau had his heart cut out—he said it was in order "to view the heart of such a wise man."

Our focus here, however, is not on cruel King Shau, but on his underling, the aging tribal Lord of Chou, who in 1143 B.C., sighed inappropriately at King Shau's rage in court. So the tyrant threw the hapless man into prison, boiled up his oldest son, and made him eat the soup. Then under a sentence of death which was delayed from day to day, the prisoner occupied his time by writing down the age-old oral tradition of the 64 hexagrams, titles, and their kernels of basic meaning. Says Legge, "I like to think of the lord of Kau [Chou], when incarcerated in Yu-li, with the 64 figures arranged before him. Each hexagram assumed a mystic meaning, and glowed with a deep significance. He made it tell him of the qualities of various objects of nature, or of the principles of human society, or of the condition, actual and possible, of the kingdom." An amazing man. He must have turned to something profound to ease the rancor in his heart.

Thus came the first known occasion of writing down all the 64 hexagrams with their titles, mathematical line structure, and basic philosophic meanings. It was a short compilation, just a statement of what each hexagram signified, no thickly annotated *Book of Changes* such as you see now. Much was added by the Confucian scholars who turned it into a self-serving treatise that more and more began to counter the original Taoist-style grit and shadow.

After two years our scholarly hero was released from prison. Eventually he even became the founder of the Chou dynasty which overthrew the Shang. But he got his kingship posthumously, twenty-two years after his death. (Perhaps it's rather like being named a saint long after one is already dead.) It happened because his son Wu was able to defeat the Shang and turn his own tribe into the founders of the new Chou dynasty. This dutiful son Wu retroactively designated his dead father to be the first ruler, as King Wen— meaning the cultivated or literary king. He took for himself the name of King Wu—meaning the warrior or military king.

Wu was not so repressive as the former Shang rulers. He adopted a reconciling policy to supporters of the defeated Shang. He allowed many subjects to keep the land they held from their former rulers, but placing his own people in protective cordons on new fiefs in the empty land around and between them. Greg Whincup reports that, "Even the Shang crown prince . . . was given land so that he could continue to propitiate the Fathers and Mothers."

Today we know that history is rewritten to cast a favorable light on whomever has come into power. Some say King Wen probably really had been plotting to overthrow the tyrant and was discovered at it, and that's why Shau put him into prison. At any rate, King Wen became a great folk hero, a good and prudent man according to historians of that period. James Legge summarizes the view of Chinese scholars and also the peasant lore when he says of Wen, "Equally distinguished in peace and war, a model of all that was good and attractive, he conducted himself with remarkable wisdom and self-restraint." Certainly the old folk song at the beginning of this section extols his merit.

King Wen's original work gives us the first layer of the hexagram dynamics. For Westerners, Wilhelm does a good job of presenting this basic meaning. But for most of the changing lines, I actually prefer Legge's version (despite his debunking tone in the commentary), for he is so ferociously exact. This second layer, the changing lines, was added by Wen's second son, who became the Duke of Chou. He elaborated on his father's work by adding the text that conveyed the meaning of each individual line. What Wilhelm calls "The Image" is yet another layer inserted by later commentators.

The old Chou tribe's sequence of hexagrams has become the standard used in casting the oracle. Antiquity called it the dynamic or mutating order whereby changing lines of a hexagram can lead into a second hexagram. This movement is based on the Lo Shu. Some scholars have said, "Why, the Lo Shu is just the magic square of 15, well-known around the world." But I think there is something numinous in the magic square's fascination. It has attracted awe in cultures worldwide, from Greece to Africa to the American colonial Ben Franklin. It calls to the blood and bones and instinct, I think, because this heartbeat of meshing number holds embedded in it the paradigm of co-chaos. We'll see it later in "The Atomic Map."

This Chou I or Book of the Chou Dynasty shows the hexagram arrangement that we moderns call the I Ching. It animates that static Old Family order of hexagrams and allows it to change. Admittedly, the Old Family order had its points, notably its great regular beauty. Shao Yung [1011–1070 A.D.] turned it into an elegant mandala, the square-within-a-circle which appears in some I Ching translations— I like James Legge's diagram, shown in Plate II following. It is a vast static layout, like an architectural plan or a machine design.

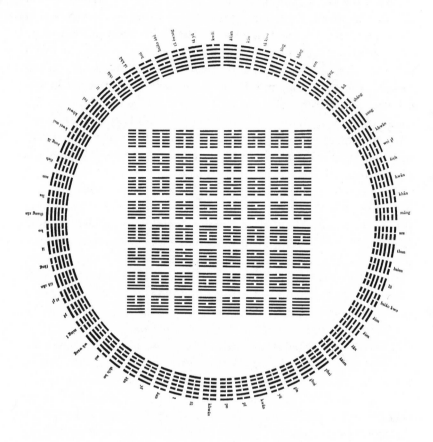

Shao Yung's Old Family Mandala

But this static layout of Shao Yung's mandala above does not describe the process of change, just as a blueprint does not manifest the traffic count in a building after it goes into use. Opposite is the *dynamic* order of the 64 hexagrams from Table I in Legge's translation. Hexagram 1 starts in the upper right corner, with Hexagram 2 to its left and so on. Sometimes Western versions will reverse this right-to-left sequence into a mirror image—I suppose because it's how Westerners read and write. Traditionally the East liked to move from the logical right side ever leftward into wordless rhythmic holism, just as we Westerners strive to push continually rightward toward a sharp, discriminating, exact edge of linear consciousness.

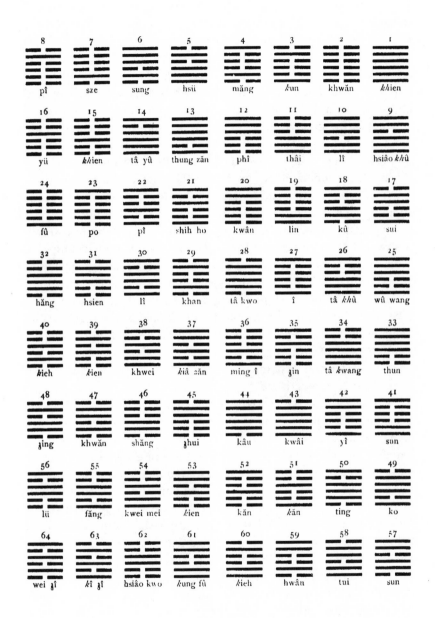

Chou I Order

People sometimes dispute which hexagram order really came first. Some also say that all these various orders merely rearrange the same basic 64 hexagrams into various sequences. But there is more to sequencing than meets the eye. We've already seen how the Old Family and New Family actually trace out different polar dynamics. For instance, you can count straight binary in the Old Family, but not in the New. The New Family has a more syncopated and subtle beat. Well, hexagram sequencing matters too. Soon we'll see that in the genetic code, too, there is more to sequencing than meets the eye. In any kind of coding, the position of a symbol can matter as much as the symbol itself—in fact, its position can convey most of the symbolism. For example, 69 is quite a different proposition from 96 or ♋, the astrological symbol for Cancer. And 8 means something entirely different from ∞, the symbol of infinity.

If yu cast a hexagram and then look it up in the I Ching, you will discover that the upper and lower trigrams really are working sub-units of the hexagram. To find ䷙ on the master chart, for instance, along the margins you must first separately locate the lower trigram of Heaven ☰ and the upper trigram of Mountain ☶. The place where they cross-grid on the chart gives you Hexagram 26, *Taming Power of the Great*. It's much like looking for a street on the map at location C4. Then you turn to the text to read about this hexagram's specific archetypal dynamic, which you then apply by qualitative analogy to your own issue.

Your hexagram will show a fractal pattern occurring in your events and emotions. It describes this energy in images that seem archaic, but the underlying pattern is enduring, archetypal. Remember that an archetypal form remains the same while its contents vary. Much like the Lorenz attractor in dynamic chaos, the iterating dynamic of your hexagram is tracing out points in a pattern that will never exactly repeat itself and whose dots can never be predicted in their exact location, yet what keeps reappearing is the same form behind the varying contents. Here is archetypal order beneath apparent randomness. It will translate into your life quite shockingly well, if you can understand the ancient example. Sometimes this is difficult, I admit, so in Section 4, I offer two hexagrams in present-day terms. I also suggest a simple, effective consulting procedure that is faster than the relatively modernday coins (1000 A.D.), and yet as accurate as the truly ancient and slow yarrow stalk procedure.

Your hexagram dynamic will tell you the quality of an event, not its quantity. Its emphasis will be on process, not final product. When the I Ching speaks of "good fortune," it considers the state of your psyche, not a prestigious accumulation of wealth. All this can puzzle the Western mind that tends to think in terms of a binary yes-no, goal-oriented, bottom-line summation of material gain.

But the East intuited this number structure resting deep in the web of nature, and then incredibly enough, used it to explore a process-oriented philosophy and psychology. The I Ching plumbs the mind, charting its hidden currents and treasures and chasms—in other words, its analinear dynamics. Meanwhile the West eventually came to an independent discovery of this same mathematical structure, but now as a statement in flesh. Biologists employed linear logic to discover these same 64 paired triplets, but as DNA spiraling with the genetic code for building blocks of organic matter.

New analinear scope is bringing us back into relationship with ourselves. We have begun to realize that we cannot get rid of a polar tension by ignoring half of it and sending it packing off into the unconscious. Disowned, it will keep erupting in strangely hostile and alienated ways. Denied in the psyche, it will manifest in alienated matter . . . even as static on the telephone wires, as Benoit Mandelbrot discovered for IBM. He was hired to find out what was causing static in the lines that ran information among some IBM computers—why they were troubled by so much "random noise." The old cure had been to try to drown out the noise by increasing the strength of the signal. That was a very yang approach. Overpower it with strength. Blot it out. Win over it. Conquer or be conquered.

But Mandelbrot discovered that the transmission static carried hidden fractal relationships within it, cycling patterns both large and small in a dynamic that was much like a fractal Cantor set in space, but this pattern was in time. Or rather, in timing. So Mandelbrot worked out a way to deal with this static by developing a new relationship to it—essentially by making friends with it and turning the analog realm into an ally. How? By cycling some redundancy into the information string to reiterate and reinforce the desired message instead of trying to oust the static in a single-minded linear battle to the death. Wisely Mandelbrot saw that merely upping the power would only up the noise too. We are learning it is better to work with and transform analog shadow than deny it.

Oh yes, the analog realm will out. It will murder our best-laid linear plans if we demonize and alienate it. It will sneak up on our blind side where consciousness doesn't shine and attack us with a shitty dark truth. If our ego boundaries are too proudly rigid and defended by glittering shards of crystalline facts arrayed to hold the dark demons at bay, then the wordless truth of the analog realm will turn hostile and try to breach our ego walls and overwhelm and destroy its brilliant identity after the sun goes down. Here are born the recurring nightmares, the intransigent domestic battles, the horrors of generational blood feuds and all-out war.

The more that mere ego consciousness is exalted as the only way, the more the analog domain becomes an antagonist . . . an artful, dangerous enemy to be feverishly repressed. The evil empire. We make monsters out of whatever is cast beyond our ego walls and left incommunicado. Linear logic insists on a battle between good and evil, between right and wrong, between a rigid righteous right way and a sinister left-handed path of darkness.

The West hasn't been very kind or gentle to those in its linear society who vibrate to a more analog reality. Those who tap deeply into this power below logic's ability to explain have generally taken on roles in the culture as the oddballs, the castouts, the rejects. Oh, eccentric analog genius has been occasionally embraced as admirable or lovable—for instance, that oddball Einstein. He couldn't add basic math in school, but he could discover the theory of relativity which changed 20th-century thought and gradually made us all more accepting of the new subjective stance of science.

To an extremely linear mindset, the analog domain holds those off-beat, crazy artists who starve for their art and no doubt deserve to, the mystics who closet themselves into isolation and out of touch with social reality. Of course, there are also the criminals who are stuffed into cells emitting the dark roar of an ethically hollow society, the insane also locked away because they are too tender or tough for this world, the homosexuals who insist on acting out the contra-sexual element inside all of our psyches but sometimes with an exaggerated, desperate emphasis that stuns the culture back into un-easy distancing, the David Koreshes with their funeral pyres of devotees . . . and then of course there are the perversely occult who gloat in their terrorizing art and debase it into orgies, blood sacrifices, and the psychological power plays of dark witches or warlocks.

Indeed, this analog domain can be frightening, for it goes into the dark side of the psyche where reason cannot go, to the very wellspring of mind itself. Logic has often labeled those who gather in the liminal twilight bordering the dark unconscious as deviates on the fringes of society, as romantics and degenerates and witches and spellbinders and dangerous lunatics escaped from the linear highways of traditionally sanctioned thought. Yet these people can enrich us even as they are despised for going beyond the pale and over the linear edge. They stride into the boundless nightways of creation where darkness bestows creative riches and hidden truth.

Seers and artists and shamans could be an integral part of our society. Dark resonance could be acknowledged and discussed and harmonized within the culture, and in oneself. Dreams could be understood and heeded. Drugs could be dropped for the true high of a meaningful life that is not just touched by god, but nestled in god.

By using the I Ching, we can do more than carry on a one-way prayer to god. Divine plan can dialog with us, can counsel us, comfort us, mystify us, nourish us. We may finally even realize that we cannot keep the analog dark under control by fighting it. Instead of befiending the shadow, we can befriend it and transform its mystery into a mine of deep riches. We can turn the lead of despair into the true gold of divine meaning in the alchemy of the soul. It is no accident that the adjective *divine* and the verb *to divine* have the same Latin root in god.

Applying the colorful imagery of the I Ching will broaden your spectrum of insight. For instance, a friend might ask you, "How was your day? Okay?" and your mind will no longer respond in a pat binary summation of yes or no, good or bad. You will not see life in the monochromatic flatness of crude black or white, shitty or swell, but rather, in the greater tint and tone of analogies. You will sense—whether you say it or not—that today went perhaps more like this: "Oh, today was like hearing thunder in the high mountains. Every rocky wall and canyon and sharp peak of detail kept echoing back to me the same shockingly powerful message, resounding it louder and louder from every event: 'Pay attention, pay attention to the details!' So I was quite conscientious today, and it has paid off." What you perceived about your day places the trigram of ☳ *Thunder* over the trigram of *Mountain* ☶ to form the co-chaos pattern of Hexagram 62, *Conscientious Attention to Detail* ䷽.

Thus, rather than a simple dualistic opposition of black versus white or good versus evil, the I Ching develops nuance through its bifurcation into eight trigrams that are mathematically equivalent to the eight possible Period 3 windows. These trigrams or windows then interact in pairs of relationship to create the 64 hexagrams of co-chaos. They describe 64 different archetypal dynamics that iterate their self-similar patterns in spacing and timing on scales large and small, moving your matter through events in all sorts of places.

It humbles me that people somehow intuited this huge order so long ago. I doubt that it was just a lucky accident. I think these ancient people tapped into cosmic pattern hidden deep in the universal structure of number because they were attuned to its voice in their creative unconscious. It resides in all our bodies, in our minds. Our minds are a subset of universal mind, just as our bodies are a subset of universal matter. I also suspect this is why the Chinese have traditionally honored their ancestors so deeply . . . it became a blurry, handed-down, generalized respect for that ancestral tapping into a truth beyond words, beyond mortal concepts of time and space.

The I Ching was passed down as a structure and philosophy of hexagrams without much technical explanation about how the polarities mesh at the corners of planes, squares and cubes into a hypercube whose cycling rhythm iterates the cosmic heart beat. I have however seen fascinating maps of ancient Taoistic rituals that in effect show a scientific approach to this knowledge; it is embedded in the procedures. It seems to me, for instance, that the ritual of the "one-footed paces of Yu" from 4,000 years ago is a description of the process of polarizing change across the hypercube formed by the intersection of space and time, matter and energy.

In Western science we are only beginning to touch the barest edge of understanding this elaborate polarized latticework. How? We make laboratory study of any polarized matter that will obligingly "seize up" enough to exhibit its lattices frozen into hard crystals or metals or the impossibly slow liquid flow of glass. This very crude level of studying lattices with "seized-up polarity" is necessary at our early stage of development.

And even this much is quite difficult for us. We enter a new world where all becomes relative, depending on what surrounds it. For example, the study of spin glass lattices was first approached by adding a few stray atoms of foreign matter to a perfect crystal. But such fastidious work proved to be a hard task, almost impossible. Daniel Stein described it as "attempting to study a clean mud puddle." Is there an easier way to study lattices? I believe that Western science is currently reinventing the wheel with patterned chaos and its vocabulary. Often in the scientific literature I come across diagrams and expositions on chaos dynamics that might be better described by the more succinct yet more accurate and versatile symbols of the I Ching.

To study co-chaos as it exhibits in the I Ching and in the genetic code weaves the past and the future together. And why not? All finally meshes into self-similar patterns on scales small and large, repeating their form but evolving their contents slowly to generate more and more complex levels of organization.

Cultures worldwide agree that there is finally some primal root where everything becomes one. Unity. God. The Tao in ancient China. The I Ching was so fascinating to its Taoist priest-scientists precisely because it showed how the dynamic play of the universe delivered a code of ethics to be found within nature itself. What nonsense, a Westerner may scoff. Some natural oracle to guide us through this maze called life? Incredible! Impossible!

Yet in 17th century China, Jesuit missionaries started a Figurist movement based on the mathematical "figures" of the I Ching trigrams. They wrote that all mathematics is rooted in the I Ching permutations. They also wrote that if this Chinese Classic were considered in its "genuine purity" established by China's ancients, free from the many insertions of more modern commentators who have turned it from the short *Changes* into an extensive *Book of Changes*, the result would be identical to the law of nature. They even tried to convince Emperor Kang Hsi that the *Bible* really was based on the same natural laws as those in the I Ching. He didn't buy it.

People even today wonder if a primal law may exist. Is there any natural basis to morality? In the preface to "The Genetic Code: Arbitrary?" Douglas Hofstadter puts the question this way: "Can cooperation and even a seeming morality emerge purely as a consequence of the laws that govern self-replication and the universe's impersonal preference for various states?" But why not consider that question conversely? Can self-replication and the universe's preference for various states emerge purely as a consequence of the laws that govern cooperation and even morality?

One possibility is actually as likely as the other. In fact, both go hand in hand, for the binary duel to the death that the West insists on reveals its own bias toward a linear, yes-and-no, goal-oriented, materialistic domain honored by current techno-science-bureaucracy and favored by the left brain. It tends to downplay the process-oriented, holistic right brain—that realm so familiar to the artists, philosophers, inventors, mystics. But obviously the universe around us contains and meshes both systems within its greater whole.

Leibniz thought there was something quite potent in the I Ching system. A devout Christian, he spent a lot of time trying to find some way of connecting Christian theology to the I Ching philosophy and binary code, but he died before anything substantial came of it. And a lot of people were relieved about that, because they thought he'd gone off the deep end. The Jesuits were eventually forbidden to continue their mission to China in 1742, and then in 1773 were suppressed by Pope Clement XIV, who decided that the Jesuits were much too interested in integrating scientific and mystic thought. Ironically enough, during the very bloom of Enlightenment, many decided that Leibniz and those heretical Jesuits had become quite irrational over number.

Getting irrational over number has proved dangerous before. To stretch a pun backward over two thousand years, irrational number was the downfall of the Pythagoreans. In ancient Greece, irrational number caused such an internal schism that many were ostracized and martyred for their math-cum-spiritual beliefs.

So as we journey into the genetic code and its merging path with the I Ching, let us go careful and rational and logical and even binary at first. Later on, we'll turn wild and crazy and drop off the linear edge into the patterning wilds of art and ethics . . . but not at first. That would be too heady and intoxicating a way to start. So we will tread a linear path along this analog maze, unwind a thread of continuity through its pages into the organizing center, that still small space in the heart of darkness.

In this space shimmers utter light. It comes through a Period 3 window that meets another Period 3 window as they shift through the possible permutations to form the 64 patterns of complementary chaos. Does this seem too simple a paradigm to be likely? Does it sound too much like a pipe dream, bubble-headed and daft, but in the French put-down phrase, how lovely to think so?

Well, remember that the best science is simple and clean and beautiful. Newton knew it when he said, "Truth is ever to be found in simplicity, and not in the multiplicity and confusion of things." Within co-chaos lies simplicity. Patterned chaos is generated from simple rules into bifurcating cascades of designs on scales large and small, repeating in spacetime through mattergy. It is all engendered from this primal archetype of number.

As I complete this chapter, last night I had a dream, most of which I cannot even remember now. But it was a lesson in life's elementary math. And fortunately, at the very end there was a little summary showing how linear and analog number join together and engender everything. Viewing this summary in the dream, I said, "Oh, look. I see. Now I'm not frightened. Father, you are the prime mover delineating my path. And Mother, you're the double bell ringing me into life and resonating me home."

And waking up, I saw that my life is my ongoing adventure in analinear play, and so are all our lives, in some liminal domain where work and play become the same thing and it all leads us to the same conclusion and beyond.

If the system of co-chaos is so simple, why haven't we noticed it before? Well, as you can see, people did, a long time ago. It is recorded in the cryptic symbols of the I Ching. It uses father yang and mother yin in either-or *plus* both-and style. Together they generate the patterns that birth the cosmos.

WORKSHOP 7

Plan a program that combines discussion and experiences. Select book paragraphs to introduce the following topics—or your own. Opening ritual.

❧ In a group, name all the movies, TV shows, books, etc. you can think of that use time travel as a motif. What is the value in this plot device? If you could travel to one date in the past or future and return, what date would it be? Why? What gain comes of it?

❧ Consider a "bad time" in your life. Perhaps you can transform the memory by befriending it as an ally interested in your growth. How did it nudge you into developing new skills or tactics?

❧ If you had to live in a cold stone prison for two years, how would you occupy your time? No legal recourse, no visitors, no leaving the cell, only 100 sheets of paper, bamboo pen, and ink. What resources could you tap within your own body and mind? Could you manage to live the time constructively? How would it change you? How would you feel about things when you came out?

Meditation: We are all locked in the prison of our own minds and bodies. Consider your own life in its solitary aspects. Your mission here: to make this time ennobling rather than degrading. How can you find ways to communicate beyond these subtle walls? How can you enrich and enjoy your sentence locked in flesh? How can you make this time constructive and satisfying, so that you will emerge from prison better instead of worse? Imagine that after your lifetime, you become released back into cosmic union with the oversoul, into the sublime ecstasy of all-union. The unique texture of your lifetime will go back into the pool to deepen it. What are you bringing to the pool from your mission? Now, before the release date, ask the beloved oversoul to help you enhance the texture of your life and improve your gift.

❧ Share a communion using chunks of yeast bread and apple juice. The yeast bread represents analog exponential growth, and the apple juice represents pressing into good service the tricky apples of linear knowledge.

Closing ritual. Five minutes of feedback. Announcements.

The Master Plan

The genetic code is a dazzler. It encodes the life plan inherited from the past, but then decodes it with unique new variations that give one person brown eyes and dark hair and a high I.Q., while another gets green eyes and red hair and a gift for music—and these two very different people can be siblings in the same family.

The big question for modern geneticists became, "How do these hereditary-cum-evolutionary instructions operate? How do they get encoded and decoded and translated into matter? The West quested to crack this code. What was its message, what was its medium?

The medium is molecules. The code is written in four different molecules—all base nucleotides, or bases for short. They are Uracil, Cytosine, Guanine, and Adenine—for short, U, C, G, and A. The first two have similarities that make them a pair of pyrimidines, while the other two are a pair of purines. These bases of U, C, G, and A can be arranged in any combo of triplets—for example, AGU or CAG or UUU or GAA. These triplets join in an amazing yet simple way that creates the DNA double helix.

DNA's double helix is a model for the controlled evolution that is possible through co-chaos. It combines the circle's repetition with the line's directional thrust to make a spiraling ladder. This twisting DNA ladder echoes the simpler Lorenz attractor, which is a looping figure 8 of iterating points that never quite exactly repeat their old locations. But DNA has an sophisticated difference: it pushes into a higher complexity by going not just over the same ground in new variations, but by evolving the "points" which are the organisms in its living system forward in spacetime.

Genetic code is found in every living thing, warm-blooded or cold-blooded or bloodless. In a virus. Amoeba. Twig. Moth. Raccoon. Fish. Human. Everything that's alive. In humans, the double spiral of DNA is found in every cell except the red blood cells.

More specifically, in nucleated cells the tiny double spiral of DNA intertwines in a way that's been compared to two interlocking bedsprings. Or to a twisty ladder, where the bases bond to make rungs. Or to a zipper, where the bases become interlocking teeth. They form a zipper, yes, but not in designer jeans—in designer genes.

In the picture on the next page you can see the genetic code at work. First comes the double helix of DNA. Its two spirals are genetic partners—black and white. The twists of their ladder-like rungs occasionally obscure a letter, but you can still guess it anyway from the letter that's visible on the partner. How so? Well, because T always partners A, while C always partners G.

The bases bond across the double helix by mutual attraction. So one spiral's sequence of bases will totally define the sequence that's found on the other. They are in complete and total complementary accord. This seals the double spiral into a fail-safe device of cross-referenced genetic information, and it stabilizes the double helix to hold the master plan intact. This DNA contains the inherited gene data. It must protect that plan carried in every egg or sperm. So DNA is stable, sturdy, a plan . . . much like the stable Old Family of trigrams. Life will be constructed when this plan gets decoded and built. That will be RNA's job, building from the plan.

But how is RNA born? Sometimes the parental DNA unzips its double spiral to procreate. The center picture shows an unzipped white strand of DNA spawning a gray RNA daughter. She will mimic the black parent's message, but not with exact fidelity. Her accent is slightly off. Every time she mimics T, it comes out U. This lisp makes the U become a variant found only in gray daughter RNA, while the T version is found only in parental DNA. This parental T has one more carbon atom and two more hydrogen atoms than the U. So the upshot is, this gray daughter mimics the black parent's information more succinctly, using this smaller molecule form called U.

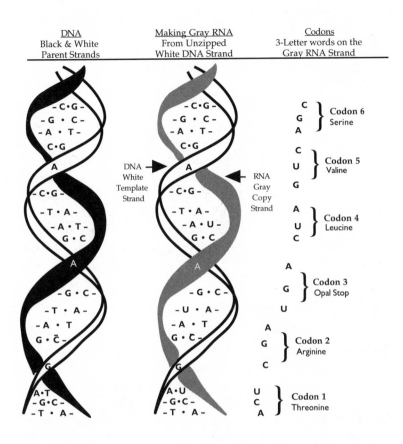

And here is the crux. Here is what makes the daughter a solitary wanderer instead of a couple like her parents, turns her into RNA instead of DNA. Her mimic U makes too fragile a bond to be stable enough for the sturdy zipper-like construction that the double helix needs to protect its blueprint of life. So RNA was born to roam.

This gray daughter will detach from the white parent, and as a solitary strand, she will carry away the building message. Where will she deliver it? To a builder ribosome. It deciphers the message by cutting the string of bases on her gray copy strand into three-letter codons. On the right is the RNA daughter's message, cut into codons. Each codon spells out a three-letter word in the building code of life.

Ribosome-Decoder

Let's look at how this happens in closer detail. Here is a builder ribosome, drawn in what's amusingly known as the two-blob model. Admittedly, this looks rather like a snail, but really it's a manufacturing plant. This ribosome makes the building material of protein. Think of it as a little portable factory which the daughter RNA threads itself through for decoding, so the factory can then build protein according to the blueprint that was sent over by the parental DNA. Here is the raw string of letters carried by this daughter's sequence:

·.·A G A C U C G A U G A C U A G U C A G C

The little ribosome factory runs along this RNA strip from left to right. It busily decodes the chain of letters by breaking it into the three-letter words called codons. Each codon spells out an amino acid or traffic signal. They connect into "sentences" of protein. Rising behind the ribosome you can see an ongoing strand of protein, made by joining the amino acids to each other in a crinkled string. All life is made of these strings of materialized code. They write the story of a living organism.

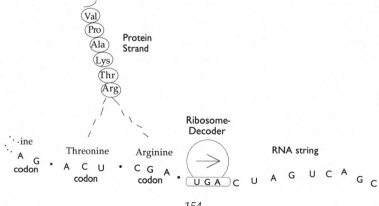

Our little ribosome factory knows that it's just about time for the whistle to blow. It will stop making protein and detach itself from the RNA strip. Why? Because that codon it's working on right now—UGA—signals Opal Stop. This translates into, "String ends here. Stop!" It is one of the three stop codons. (There is only one start codon.) Sometimes the end is double-signaled by two different stop codons coming right after each other, like amber and red traffic lights when you're driving. It's as if the code is saying to the ribosome, "Caution, get ready to stop. Now . . . really stop!" These traffic codons give the ribosome factory its stop-and-go instructions. They are similar to the instructions written into a computer program that tell it when to start and stop its data-processing.

In summary, the black and white double spiral of parental DNA holds the stable form of the genetic code intact—the building plan—while the gray RNA daughter carries the message that eventually will be materialized into protein. Gascuel and Danchin put it this way in *Data Analysis Using a Learning Program,* "Using an analogy between a living organism and a building, one may say that the DNA represents the architect's plans and proteins represent the building materials and machinery."

Stable or restless, each version has its own task. Where DNA must be stable, the daughter RNA must be mobile and flexible. She must be able to cut loose from the white parent after replication and roam around the cell as a single open-toothed messenger instead of a closed double zipper. So in RNA, the more fragile-bonding of uracil or U is necessary for any progress to develop.

What does all this have to do with the I Ching? Precisely this: now we can cross-code the genetic code with the I Ching. The result will be a hybrid I Code. It will show how both formats, East and West, actually follow the same math model. In it, a Period 3 window counterposes another Period 3 window. In the I Ching, this form is called the six-line hexagram; in the genetic code, it's the paired triplets of DNA. Complementary chaos results.

To get more nitty-gritty specific, the genetic code bifurcates into pyrimidines and purines. Next, the pyrimidines fork into Thiamine/Uracil or Cytosine, while the purines fork into Guanine or Adenine. Here's the beginning of the familiar bifurcation tree! Of course, the Tao also divides into yin and yang, and next, yin forks into stable or changing yin; likewise with yang. In each version, we can see the characteristic polarized forking of chaos patterning.

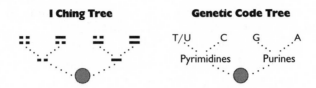

I Ching Tree   **Genetic Code Tree**

Let's crossbreed these two charts into a hybrid I Code tree. Here stable yin equates to T/U, changing yin to C, changing yang to G, and stable yang to A:

I Code Tree

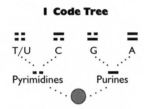

Consider our I Code tree rising in the center of the fractal garden. Perhaps you think that any other match between letter and line would serve equally well. But no, this is the only match that really pans out. You'll see why as we go along.

This tree has forked upward into four branches. Can it reach a still higher level of shared order? The next level up for the I Ching would be the trigrams, each with a trio of lines. Hmm, a codon also has a trio . . . of bases!

Okay, the genetic code has 64 codons. And the I Ching has 64 hexagrams. So they really do echo each other. But are they really parallel? Take a look at the chart on the next page. On its left are the hexagrams. On its right are the codons, just the way they are listed in a typical biology textbook. We can see that each has 64 units—64 hexagrams on the left and 64 codons on the right.

But there's a problem. A hexagram is made of *two* trios: a pair of trigrams with three lines each. But we see here only one trio for a codon—for example, underlined on the chart you can see the four traffic codons of AUG, UAA, UAG, and UGA. Each has three letters.

Old Family Hexagrams

Genetic Codons

UUU	UCU	UAU	UGU
UUC	UCC	UAC	UGC
UUA	UCA	**UAA**	**UGA**
UUG	UCG	**UAG**	UGG
CUU	CCU	CAU	CGU
CUC	CCC	CAC	CGC
CUA	CCA	CAA	CGA
CUG	CCG	CAG	CGG
AUU	ACU	AAU	AGU
AUC	ACC	AAC	AGC
AUA	ACA	AAA	AGA
AUG	ACG	AAG	AGG
GUU	GCU	GAU	GGU
GUC	GCC	GAC	GGC
GUA	GCA	GAA	GGA
GUG	GCG	GAG	GGG

U = Uracil   A = Adenine

C = Cytosine   G = Guanine

What could turn a codon's threesome into a hexagram's sixsome? To exhibit a shared plan, they must follow the same structure. In each of the 64 groups, a triplet must counterpose another triplet. Have we lost our I Code tree? No. It's hidden in the undergrowth of detail. We'll scout out a path by following clues along the way

The first clue comes when we recall that the stable Old Family comes from the Ho Tu, called the Plan or Map. However, what's necessary to activate that plan is the Lo Shu, called the Writing.

Well, a stable plan is also found in DNA. It locks the gene into double security. What makes that plan become dynamic, what makes life happen, is unlocking its stable double helix to spawn the traveling single row of RNA. This halved form is what turns the blueprint into the amino acids that build protein. Just as the Lo Su Writing materializes the Ho Tu Plan, so does RNA protein writing materialize the DNA plan. In genetics, it's even called transcription!

Here's a short swatch of DNA. Reading it upward, the white bonding strand of GCA bonds to the black strand of CGT (which will later turn into the gray daughter RNA codon of CGU). And it is here that the gene's pair of triplets emerges! The paired trios become manifest. It has the same plan as a hexagram of two trigrams. They both share the co-chaos model of bifurcating paired trios.

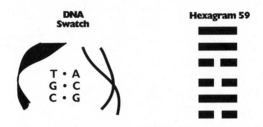

DNA Swatch

Hexagram 59

Paired Trios occur in both Genetic Code & I Ching

Now if we take our DNA swatch and number the bases, we see the first and fourth will pair-bond, as do the second and fifth, and also the third and sixth. It's called the bonding of gene base pairs.

But it also holds true for the I Ching! Amazingly enough, there is an ancient and exactly comparable rule for pairing I Ching lines within a hexagram. Quoting James Legge in his *Translator's Introduction to the I Ching*, "The lines, moreover . . . are related to one another by their position, and have their significance modified accordingly. The first line and the fourth, the second and the fifth, the third and the sixth are all correlates . . . . " Talk about shocking! This takes the shared paradigm even further. To see it in action, compare our DNA swatch with Hexagram 59, *Dissolving of Blocks*.

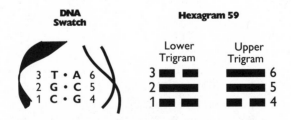

DNA Swatch

Hexagram 59

Lower Trigram   Upper Trigram

Pair-bonding occurs in both Genetic Code & I Ching

By the ancient I Ching rule, read upward from the bottom:

Line 6 ▪▪ in the upper trigram
correlates with
Line 3 ▪▪ in the lower trigram

Line 5 ━ in the upper trigram
correlates with
Line 2 ━ in the lower trigram

Line 4 ▪▪ in the upper trigram
correlates with
Line 1 ━ in the lower trigram

This I Ching rule exactly parallels the gene's pair-bonding rule. Comparing DNA swatch with hexagram, we get the same bonding pattern for both structures. The black parent equates to the lower trigram; the white parent equates to the upper trigram.

The DNA swatch has six items . . . and so does the hexagram. The swatch divides into two polarized triplets, and so does the hexagram. The swatch bonds Item 1 to Item 4, Item 2 to Item 5, and Item 3 to Item 6, and so does the hexagram. The swatch is built on increasingly higher orders of bifurcation, and so is the hexagram. And of course, both run through their 64 possible permutations. This is the same mathematical structure. It is both liner and analog, with each sixsome forming an analinear equation.

Thus the I Code theory postulates that, "The DNA swatch is two counterposed Period 3 windows of chaos patterning, and so is the I Ching hexagram. The 64 DNA swatches and the 64 hexagrams both follow the same paradigm of co-chaos. All this happens not only according to ancient Chinese rules, but also according to genetic bonding rules. Complementary chaos exists in each system, and the two systems mirror each other, one expressing itself in mind, and the other in matter." Here is the place where mind and matter finally meet. It is in co-chaos!

But wait, there's more.

Look at Position 6 of this shared plan. According to both the genetic code and the I Ching rules, this sixth position is "wobbly." It cannot be depended on to guarantee any firm direction. The gray daughter RNA version (using U instead of T) and the hexagram both exhibit this special and atypical behavior in Position 6.

Let's consider the genetic wobble first. In the ribosome, a tiny worker in Position 6 (called a tRNA) holds the least clear and reliable spot, allowing it to "wobble" or opt several different ways in decoding. This unreliability can allow several different codons to make the same amino acid. For instance, CGA, CGC, CGG, and CGU all decode into Arginine, because of Position 6's wobble. Due to this wobble, these four codons wind up as the same amino acid.

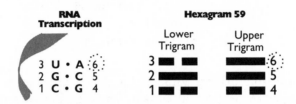

Position 6 Wobble occurs in both Genetic Code & I Ching

Now compare this genetic wobble with an old I Ching rule for the sixth line of a hexagram. It calls Line 6 the farthest removed from worldly influence, too late in the sequence to hold a strong directive. But it is "closest to heaven" and thus it carries a remote purpose that is not readily visible or effective in ordinary earthly, material terms. So it exhibits a behavior that seems unrelated or clumsy in normal terms, since it has another purpose instead. It serves a higher goal.

This description parallels the wobbly codon behavior! Seemingly slipshod, wobble instead gives a latitude that allows the codon families to form into a higher level of organization. (For this reason, and for others, I think the four-letter alphabet that genetics currently uses for codons just doesn't do justice to some of the most vital features).

Thus, wobble is yet another principle that is parallel in both the genetic code and the I Ching. In all these various ways, the I Ching and genetic code share the same structure.

But wait! If this much is so—if the 64 DNA swatches really do follow the same plan as the 64 stable hexagrams—then perhaps they correlate even further. How to take this quest onward? How can we validate this shared plan as being so far beyond chance that an accidental similarity is impossible?

How about showing that the codons and hexagrams even carry equivalent meanings? For example, that the four traffic codons really do equate to four hexagrams giving start-or-stop instructions? Or that the natural amino acid tranquilizer called Tryptophan actually equates to the calming, reassuring Hexagram 35 of *Easy Progress?* Or that the amino acid for Hexagram 14, *Possession in Great Measure,* actually lengthens the bony measurement of your skeletal structure? Well, all this does come about, and more, as we shall see.

But how can such a thing possibly be, you may wonder. How can an amino acid act like a hexagram's philosophy? Just consider the fundamental plan that we have here. It uses number, the primal archetype, and in a way that is both linear and analog. It is not just binary logic shunting in sequences of on-off; it is also analog ratios resting in nests of relationship. Together they mesh into the co-chaos paradigm, which is *both* qualitative and quantitative. So each of these 64 analinear equations will *necessarily* make a statement that has both qualitative and quantitative features.

A more familiar example of this quantity-quality play of number occurs in atoms. When you know how many protons are in a particular atom, you know which element it is. And each element is different. It will vary from all the other elements, not just in its *quantity* of protons, but also in its *quality* of characteristics. It will differ in kind, not just degree. We even speak of an elemental difference

Let's get down to the fine detail on this I Code tree. Check out every characteristic. For it to be valid, it must be a hybrid tree that merges science and philosophy. Old Family hexagrams must group DNA swatches into coherent families of amino acids. An amino acid should have physical properties that echo its hexagram's philosophy. Traffic codons should translate into equivalent messages. Every aspect should meld structure with meaning, math with philosophy. Here is the place where unity resides and yet diversity begins. It is a hybrid of East and West, old and new, mind and matter, heart and head, objective and subjective. You will see why I settled on this one particular correlation between I Ching symbols and genetic bases. It is the only one that grew the I Code tree.

But frankly, it was hard to do. I'm showing you the excursion tour of highlights. But I made a lot of wrong turns along the way through this mental jungle. For the bitter truth is that murky swamps and obstructing branches and sheer cliffs abounded. For example, is

Guanine coding for yin, changing yin, yang, or changing yang? Where are the changing lines, anyway? I don't see any on the Old Family chart. And how to correlate the gene's four bases with only two visible symbols in the Old Family hexagrams—that stable yin and stable yang. Answers to these questions gave more clues.

And wobble! Since codons trigger wobble, mere proximity on a chart does not necessarily mean that they belong to the same amino acid family. For instance, by their alphabetical sequence, it looks like CAA, CAC, CAG, and CAU make one family. But CAA and CAG form Glutamine, while CAC and CAU form Histidine. Two different families. Even this became a clue on the path to the tree. Fortunately, only one way would fit right. Testing and retesting became my sharp machete of logic through the tangles of bewildering analogs.

Actually, merging the genetic code with the hexagrams became easy once I found the hidden changing lines. Again, the key was in the shared rules. And again, it came from Legge. Recall that a trigram not only bonds itself together, but it also bonds with another trigram. Line 1 bonds to Line 4, Line 2 to Line 5, and Line 3 to Line 6. Each bond across the chasm will form a higher order bigram, a weird sort of bigram uniting the two trigrams. It pulls the hexagram together. Three ways, in fact—through Lines 1 & 4, 2 & 5, 3 & 6.

So I thought, "Maybe a clue to changing yin and yang is hidden here . . . in the bond-bigrams." Not just a clue, I realized eventually, but the secret itself. You've seen Legge's remarks on bonding, but now I add that final crucial phrase in italics: "The first line and the fourth, the second and the fifth, the third and the sixth are all correlates . . . *and to make the correlation perfect the two members of it should be lines of different qualities, one whole and the other divided.*"

What does this mean? What is a "perfect" correlation of lines? Simply this: a "perfect" pair will bond yin with yang across the trigrams. For instance, in Hexagram 11, *Harmonious Union*, which you can see way back on page 255, the lines sit in three perfect pairs. Each pair—1 & 4, 2 & 5, 3 & 6—bonds a yang line with a yin line. Each of them makes a perfect bond-bigram.

But look at Hexagram 59, *Dissolving of Blocks*, just on the opposite page. Its lines sit in a different dynamic. Here only Lines 3 & 6 are a perfect pair. The others are imperfect: Lines 2 & 5 are yang-yang, while Lines 1 & 4 are yin-yin. According to the I Ching rule, bonding like with like (yin-yin or yang-yang) is possible, but not "perfect."

Aha. Here is the final clue to the mystery. Now I'm going to show you something so profound, yet so simple, that it may slip right past you. It's like a magic trick . . . unless you're watching carefully, you don't see how it's done. This is no sleight-of-hand, though, that the I Ching performs. Instead, it is a peculiarly simple sleight of mind.

The magic resides in a simple coding trick. Usually we forget that a pair of lines bonded across trigrams is really just a huge splayed-out bigram. In this higher-order bigram, a perfect bond is yin and yang. It is stable, balanced. But an imperfect bond is all-yin or all-yang and therefore unstable, prone to change. Hidden here is the key to decoding the genetic bases into stable or changing lines.

My study suggested that since perfect pairs are stable, they would maintain a stable code, top or bottom trigram. But since imperfect pairs are unstable, they might blur their coding, top and bottom. And this hypothesis delivered the fruit of the tree!

Here's the way to decode perfect pair-bonds into gene bases: Yin will *always* symbolize T whether it's in the upper or lower trigram . . . *if and only if* its partner is yang in the other trigram. Why? . . . because that partner must always symbolize A. Together they will always make a "perfect" pair. Each attracts its mate of the other pole in the opposite trigram. Thymine–Adenine make a perfect pair, stable.

This decoding rule means that the perfect bond-partners of yin and yang are *always* T–A bonding. They are stable and dependable. By looking at any bond-bigram, we can determine, first, if it's a perfect bond. If so, then it must be Thymine ▪▪ and Adenine ▬. You can see this happening in Hexagram 59. Lines 3 and 6 form a perfect bond of ▪▪ and ▬. We can rest assured that Line 3 is Thymine and Line 6 is Adenine.

Imperfect & Perfect Bonding occurs in both Genetic Code & I Ching

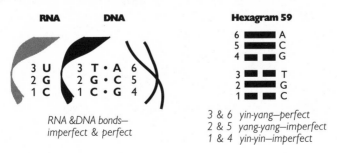

RNA &DNA bonds—
imperfect & perfect

3 & 6 yin-yang—perfect
2 & 5 yang-yang—imperfect
1 & 4 yin-yin—imperfect

But what of imperfect pairs? We've still got half of the cross-code to crack. *Here's how to decode imperfect pair-bonds into gene bases:* imperfect pairs actually change their code according to position! You may think, "But how can one read a symbol that changes its very meaning according to its position? You do it all the time, actually.

This *b* is a *d* or a *p* or a *q* depending on its position.

Position holds the final clue in this alphabet code. And it is also true for the imperfect pairs. Consider Hexagrams 48 and 59 below. In each, the perfect bond-bigrams *must* be pairings of A-T and T-A. Why? Only these bases form the stable bonds of yin with yang. But the imperfect bond-bigrams *must* be pairings of C–G or G–C. Why? Because only these bases always make imperfect bonds . . . either as yin-yin or yang-yang. Which base is which will depend on its position, whether it is in the lower or upper trigram. Thus, the line coding for a base in imperfect pairs will change, depending on its position in the hexagram.

Below are some sample hexagrams and a written guide to their different bond-bigrams. Finally, it is summarized in four code keys.

Stable and Changing Lines

Perfect Pairs—stable

DNA Swatch Hexagram 48	DNA Swatch Hexagram 59
6 ▬▬ ▬▬ T	6 ▬▬▬▬ A
5 ▬▬▬▬ C	5 ▬▬▬▬ C
4 ▬▬ ▬▬ G	4 ▬▬ ▬▬ G
3 ▬▬▬▬ A	3 ▬▬ ▬▬ T
2 ▬▬▬▬ G	2 ▬▬▬▬ G
1 ▬▬ ▬▬ C	1 ▬▬ ▬▬ C

Imperfect Pairs—changing

DNA Swatch Hexagram 48	DNA Swatch Hexagram 59
6 ▬▬ ▬▬ T	6 ▬▬▬▬ A
5 ▬▬▬▬ C	5 ▬▬▬▬ C
4 ▬▬ ▬▬ G	4 ▬▬ ▬▬ G
3 ▬▬▬▬ A	3 ▬▬ ▬▬ T
2 ▬▬▬▬ G	2 ▬▬▬▬ G
1 ▬▬ ▬▬ C	1 ▬▬ ▬▬ C

In perfect yin-yang pairs . . .

Yang is Adenine in both top & bottom trigram.

Yin is Thymine/Uracil in both top & bottom trigram.

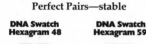

The • is the bindu point between trigrams.

In imperfect yin-yin pairs . . .

Yin is Guanine in the top trigram,

but

Yin is Cytosine in the bottom trigram!

$$\overset{\text{-- --}}{\underset{\text{-- --}}{\bullet}} = \overset{G}{\underset{C}{\bullet}}$$

In imperfect yang-yang pairs . . .

Yang is Cytosine in the top trigram,

but

Yang is Guanine in the bottom trigram!

$$\overset{\text{▬▬}}{\underset{\text{▬▬}}{\bullet}} = \overset{C}{\underset{G}{\bullet}}$$

These four keys open the way to cross-correlating the hexagrams of the Old Family chart exactly and validly with the genetic code, both by meaning and mathematical structure. Perfect bonds of yin and yang will code for T/U and A in lower and upper trigrams. They will perform true to name, with coding that stays unchanging whether in lower or upper trigram. They are stable.

But imperfect bonds are unstable, changing. In a yin-yin pair, the yin stands for Cytosine in the lower trigram, but for Guanine in the upper trigram. Likewise, in a yang-yang pair, yang stands for Guanine in the lower trigram, but for Cytosine in the upper trigram. Thus their coding flip-flops. C and G literally swap their codes between trigrams. *They are literally changing lines.*

This sort of code is entirely positional. The difference between a stable and changing line will hinge on its partner and position in the hexagram. Notice that in this coding system, you can't read a bond-bigram alone, only in relationship to its partner in the other trigram. It takes the whole hexagram to see whether a line is stable or not.

To the Chinese, this coding by position would seem a perfectly reasonable development. Look for instance at what happens when we invert the ordinary bigrams. Inverting these bigrams leaves two of them stable, but it alters two. It flip-flops the coding for C and G, which literally exchanges their lines. Yet the coding remains stable for T and A; they are unchanged whether flipped or not.

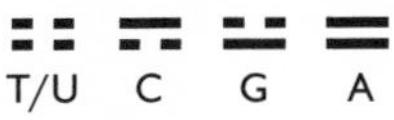
T/U C G A

Ordinary bigrams, inverted,code as . . .

T/U C G A

Moreover, ordinary bigrams hold the secret of the bond-bigrams! Adding a bindu point inside reveals how to decode them into the bases of a hexagram, both by the kind of line and by its placement.

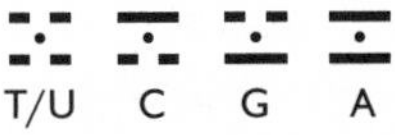
T/U C G A

Ordinary bigrams plus the bindu point become Bond-bigrams with the position code for each base!

Now we can correlate the 64 hexagrams with the 64 DNA swatches. It's easy. Easy as telling the difference between b and d and p and q.

DNA Swatch Hexagrams

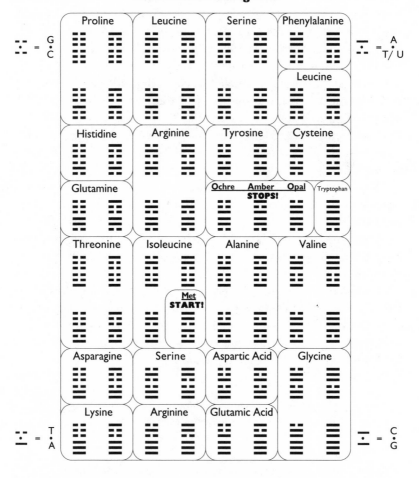

This is a Swiss army knife of a chart. Actually, we could just look at the Old Family chart of hexagrams in its simple elegance, enigmatic and harmoniously balanced, and say, "This is the genetic code." Every nuance is hidden there. But for decoding ease, I tacked the four code keys onto the corners. Here each swatch and its hexagram work interchangeably, giving textbook results—and more. It not only smoothly gives all the 64 codons, but groups them snugly according to their amino acid families. Unlike a textbook chart, it even puts the traffic codons into appropriate sequence.

In each swatch hexagram, the upper trigram shows the white parent. The lower trigram shows the black parent, which also holds implicit in it the RNA codon, simply by changing its T into a mimic U. The codon can then explode into its operant message that will build protein. Furthermore, its amino acid function will accord with the philosophic meaning of its exploded RNA hexagram. For instance, each traffic codon when exploded will deliver an appropriate stop-or-go message in the I Ching itself. In toto, it becomes evident that the same paradigm underlies both I Ching and genetic code— the paired trios of co-chaos.

These were the tests for acceptable congruence. But it took me quite awhile to administer them all. Lots of trial tables. I was a slow learner who kept considering alternate routes and resting on a bench to think it over again, double-checking each step, backtracking the false steps along the way.

In each of the four corners of the chart, you'll find a code key and its swatch hexagram *par excellence* to showcase each base pair. For example, in the upper left corner, the hexagram of ䷁ is built entirely from the C/G pair. This hexagram of ䷁ has the CCC black parent in the lower trigram, and the white parent GGG in the upper trigram. The result is all yin, top and bottom. Thus G and C form this DNA swatch hexagram out of three imperfect bond-bigrams that provide yin in every line.

Gazing across the top row of hexagrams, you see that all its lower trigrams are Black Earth ☷. But this does not mean that they all code for CCC—oh no. Since hexagrams only work through *relationship* between their trigrams, the lower line of ▬▬ is C only if its partner above is also ▬▬ (which in the upper trigram will decode as G).

The hexagram in the upper right corner is ䷏. Line 1 is ▬▬. Alone it could signal either C or T/U, but we must check its companion Line 4 in the upper trigram. Line 4 is ▬. So between trigrams, we have a perfect bond. They form a stable pair where Line 1 has to be T/U and Line 4 has to be A. This hexagram of ䷏ is formed of three such bond-bigrams. It is summarized by the key in that corner.

Code keys also decipher the other two corner hexagrams of ䷗ and ䷀. In fact, by a simple process of simple decoding, all 64 hexagrams will equate to DNA swatches in this way. And of course, you can also read all their lower trigrams as gray RNA codons, by just changing each T to U.

A beautiful ordering pattern results. It all happens because of the flip-flop coding for bond-bigrams. The two kinds of purines and pyrimidines are distributed diagonally from the four corners of the chart in even progression. The C runs from the upper left corner toward G in the lower right, while U runs from the upper right toward A in the lower left. Compare all this symmetry with the standard textbook order, and you realize suddenly that the old format has a paucity of information and harmony. We will explore more of this artful elegance in the chapter called *Art in Number and Image*.

Codons in Old Family Order

CCC CCU CUC CUU UCC UCU UUC UUU

CCA CCG CUA CUG UCA UCG UUA UUG

CAC CAU CGC CGU UAC UAU UGC UGU

CAA CAG CGA CGG <u>UAA UAG UGA</u> UGG

ACC ACU AUC AUU GCC GCU GUC GUU

ACA ACG AUA <u>AUG</u> GCA GCG GUA GUG

AAC AAU AGC AGU GAC GAU GGC GGU

AAA AAG AGA AGG GAA GAG GGA GGG

The flip-flop trait in the bond-bigrams has, I believe, been a big stumbling block in previous efforts to cross-code I Ching and genetic code in the scanty literature on the subject. First came Marie Louise von Franz's brief note in that 1968 German essay, already mentioned. Gunther Stent's *The Coming of the Golden Age* was published in 1969, and Martin Schoenberger's *The I Ching and the Genetic Code* in 1973. The January 1974 issue of *Scientific American* contained an article by Martin Gardner exploring the binary math of the I Ching. Then came Eleanor B. Morris's odd, obscure pamphlet called *Functions and Models of Modern Biochemistry in the I Ching,* published in Taipei in 1978. In 1991 came Johnson Yan's book called DNA *and the I Ching.* None of these admirable attempts at cross-coding have, I believe, given results as satisfactory as the I Code tree and co-chaos.

Biologist Stent, medical doctor Schoenberger, and chemist Yan all use different cross-coding systems. Each one considers various binary aspects, but no analog aspects. None utilizes trigrams or bond-bigrams for DNA. Each seems to get some answers that work and some that don't. Three of Schoenberger's traffic codons make sense, for instance, but calling GUG the Met-Start codon and equating it to Hexagram 4 cannot work, neither in genetics nor in the I Ching. Yan considers no tie-in between hexagram meaning and amino acid function.

But the flip-flop code of the bond-bigrams resolves such conflicts. It explains why Schoenberger had to flip half his table arbitrarily to get his codon chart to work—without being able to explain why. Eventually, there came a point when I realized that the most serviceable theory would use a flip-flop code in accord with the bond-bigram changing lines. And this, I think, has turned out to be so.

But more important still is the main issue completely unaddressed in previous literature. The crucial co-chaos paradigm has been overlooked. It offers a scientific model whereby two Period 3 windows bond into a swatch hexagram running through its 64 "nonlinear" possibilities. This gives the pivotal explanation for those myriad clues that many of us have been walking around to gaze at in wonderment.

Science does itself a disservice, in my opinion, by concentrating so exclusively on the binary aspects of the genetic code. True, most biologists see the gene as linear, quantitative, and product-oriented. True, Leibniz saw the I Ching as binary. Those binary aspects do exist, certainly, but they don't tell the whole story. Concentrating on only the linear domain deprives us of *the* major parallel between genetic code and I Ching structures—those two counterposed Period 3 windows of complementary chaos creating 64 different patterns.

Taking a traditional linear approach is intellectually hazardous in dealing with the natural world nowadays. It doesn't suffice to explain atoms or quarks or DNA. *Binary* merely indicates the 0-1 shunting of a discrete chain of logic. It discounts the integrating and transcendent properties that are inherent in analog number, and thereby it loses the complex sophistication of cycling proportions in relationship.

Going only binary, we would miss the analinear equation that forms each hexagram. We would completely overlook this merger of body with mind, of science and philosophy. But this is the whole key! By modeling its amazing combination of binary structure *plus* analog relationship, the I Code reveals to us the master plan.

WORKSHOP 8

Plan a program that combines discussion and experiences. Select book paragraphs to introduce the following topics—or your own.

Opening ritual.

❣ What are the dangers of seeing life in only a logical or only an analog way? Cite examples. How would you like to reform the scientific and arts communities to engender a better everyday world? What can you do to engender it in your own daily life?

❣ *Making DNA:* Stand up and count off by U, C, G, A. Now, U and G, raise your *left* hand: U, extend your little finger, and G, extend your thumb. But C and A, raise your *right* hand: C, extend your little finger and A, extend your thumb. Now wander around in the organism cell (the room) to find someone extending the opposite hand and finger to your own. Join these thumbs and little fingers to make a one-armed bridge overhead. Line up in a chain of one-armed bridges. With your free hand, hook onto the person beside you. You are now a bonded double helix of DNA. To background music, spiral round the room in the double-conga line of DNA. Come to a halt and assess what you've noticed.

Making RNA: In the same double row, choose which side is the black mother or white father. On the mother's side, each T will drop its one-armed bridge and tell its row neighbors, "Drop your bridge and come with me!" as it turns into wandering U and pulls away from the double row. This disbonding makes gray daughter RNA. It has no bridges, only a single conga line that wanders in the room looking for a ribosome to read its message.

Reading codons: Regroup the white parent members into a ribosome factory. Thread the single line of RNA through its factory. It must break the line into threesomes and give them useful names.

❣ Devise a pair-bonding dance to depict a DNA swatch and its attributes, using six group members. Perform and then discuss it.

❣ Mediation: Motion and emotion. How do they connect in you? How can you modify your motion to create more satisfying emotion? How can you combine plan with action in an effective way?

Closing ritual. Five minutes of feedback. Announcements.

The Master Builder

Codons build flesh. An overlay code "explodes" the DNA black parent into its gray daughter RNA message, and you and I get built.

What? Another code on top of the old? Indeed, yes. The genes are amazing in that, like in spy messages, one code can overlay another. Each subcode handles only one specific step in the larger building process. The result is a deep fractal pattern of codings, with subcodes that are subsets of the original pattern. For instance, biologists have found that genes can lay one message on top of another in a sort of offset process along the spiral. And the tiny tRNA anticodon uses a subcode where both its bonding pairs and its position form parts of the code. Why would nature do this? Because such coding is a very efficient way to save space and energy.

The easiest overlay technique is simple: a ribosome can read a string of bases and break them into three-letter words starting at a certain point—and create protein from this. Then it can read the same string of bases again, but starting at a new spot, and break it into different three letter words that code for different proteins. This occurs merely by moving the starting point for decoding up or down one space. The result gives a whole different array of proteins. Any message will depend on just what starting point you choose to decode the string.

For example, in this string of UGGACGUCCAGUUGACU, the ribosome can find three different building messages, depending on where it starts breaking the string apart into three-letter words:

UGG-ACG-UCC-AGU-UGA-CU .

. UG-GAC-GUC-CAG-UUG-ACU

. . U-GGA-CGU-CCA-GUU-GAC-U . .

In some places in the gene, a message apparently can be read only one way to make sense and translate into protein. But other places can lay one code on top of another in an offset process.

To see how the I Ching reveals one code within another, we move from the DNA swatch to its mimic offspring, RNA. We go from master plan to master builder. Basically, the procedure applies the same code that we've already used, but now only to the lower trigram. It explodes a lower trigram into its full-blown message.

To do it, we will use the line symbolism favored by Schoenberger in his book, *The I Ching and the Genetic Code*, but we will use it in a different fashion. Schoenberger used it because he saw it *worked*, without being able to explain why, and for him, it was also necessary to flip half his symbolism into G-A instead of A-G. He said that the results " . . . add up to a phenomenon which simply cannot be argued away." Although lacking a logical justification to flip only half the symbolism, he saw that it somehow made the whole chart viable. And he admitted that his method was provisional: "All the same, this must be regarded as a provisional experiment. As the arabic numbers in the margin show, this sequence does not reveal a mathematical order . . . . However, if we turn the sequence A-G round to give G-A without, so far as I can see, any disturbance necessarily occurring in the arrangement, (since, of course, the A-G sequence also seems to be arbitrarily selected), there emerges—all at once—a precise mathematical ordering of the whole . . . ."

Yes, flipping A-G into G-A was arbitrary for Schoenberger. The arbitrary reversal of half of one's symbolism is unsound. It's rather like saying that it's okay to reverse half the letters in the word ARTS, because you still get a word—RATS. True enough, the half-flipped version also has a meaning, but it's a *very* different meaning. But which is the "right and proper" version of these letters? Logically, such a question shows the double-bind syndrome that is simply inherent to *either-or* thinking—either ARTS or RATS. But analog thinking embraces multiple options. It sees ARTS and RATS and also STAR turned backwards. It was notions such as these that spurred me on to seek some common ground for the many different possibilities. It turned out to be co-chaos.

Nothing in our I Code needs to be flipped arbitrarily. Instead, it flows naturally and easily because it sits on the fundament of co-chaos. Since it is both analog and linear, it all fits. It writes all the gray daughter codons into correct amino acid families and philosophies effortlessly. It also allows us to distinguish between two layers of coding—that of the DNA plan holding the genetic code locked in stable security, and that of the RNA builder, when the *transcription* of RNA moves to *translate* the amino acids into protein. The terminology that is italicized here is used in genetics; amazingly, it even parallels verbally the old I Ching's distinction between the Ho Tu Plan and the Lo Shu Writing.

To apply the RNA subcode, remember that it is the builder. Thus it will put change into effect as T becomes U in the lower trigram and turns into gray daughter codon. Amplifying the stable lines of U and A merely makes them doubled and still stable. But amplifying the changing lines of C and G show their changing quality. Thus this subcode amplifies the three lines of the lower trigram to expand it into its full message. For example, the lower trigram of DNA Swatch Hexagram 60 is AGC. Now let's explode that into its full-blown message. Easy enough. We see that the bottom line is part of a perfect pair. Stable A — in Line 1 thus expands to ═, and we have the bottom two lines of our message.

Hmm, Lines 2 and 5 belong to an imperfect pair, changing. So we amplify Line 2's G from — to ▬▬. This gives us the next pair of lines in our rising message of ☵.

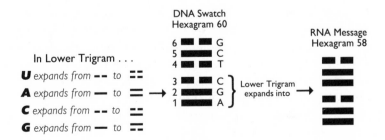

To get the RNA hexagram, expand the DNA swatch lower trigram.

The final bond-bigram is also imperfect, but in a different way. So we show this as Line 3's C amplifies from ▪ ▪ to ▬▪. When all is said and done, we have six lines that reveal AGC's message to the ribosome. This message is both quantitative and qualitative; it is Serine and also the Chinese trigrams of Lake over Lake, or Hexagram 58 ䷹ of *Shared Joy*. Its dynamic is vitalizing success through shared joy. Wilhelm's translation says, "True joy, therefore, rests on firmness and strength within, manifesting itself outwardly as yielding and gentle . . . . When two lakes are joined they do not dry up so readily, for one replenishes the other."

How do we analyze Serine for this quality of *Shared Joy* ? Why not look at the way Serine functions in the body? Serine helps the brain and nerves to function properly. A lack causes laborious thought and unsteady nerves; the whole central nervous system does not work well together. Judge for yourself whether you can see a parallel between Serine's function and its hexagram meaning. This is always a subjective, qualitative judgment.

In a strange way, analog associations are never wrong; they are just nearer or farther from the central truth, as though accuracy moves in resonant rings of propinquity that can get warped when too far from the center. To notice analogs and heed them accurately, one must true up the inner gyroscope and walk into the echoing halls of intuition toward a central truth along the resonant labyrinth of the right brain hologram. Seeking truth in here is scary for those who shun dreams, neglect their dark side, disown shadow. But the analog domain can become rich and beautiful, a place of inspiration and creation. Its analog richness, however, should always be examined and paralleled against the left brain's logic, to see whether the whole is split and divided against itself . . . or if its view of reality is congruent and harmonious.

In the web of analogs where no one can prove me wrong—or right—I intuit a correlation between Serine and *Shared Joy,* In fact, it resonates as the phrase "Laughter is the best medicine." You may protest, "But how can a qualitative correlation be proved scientifically?" Well, here is that nebulous place, for example, where Chinese medicine works. From personal experience I know that it works quite well indeed . . . for some conditions. Others, however, respond better to Western medicine.

RNA Message Hexagrams

Proline	Leucine	Serine	Phenylalanine

Histidine	Arginine	Tyrosine	Cysteine
			Leucine

Glutamine		**Ochre Amber Opal STOPS!**	Tryptophan

Threonine	Isoleucine	Alanine	Valine
	Met START!		

Asparagine	Serine	Aspartic Acid	Glycine

Lysine	Arginine	Glutamic Acid	

It is easiest to see a clear correlation between traffic codons and their hexagrams. Consider the three Stop Codons which you find underlined and boxed together near the center of the chart. The Ochre Stop amplifies into Hexagram 33 ䷠ *Retreat*. It is a traffic signal for the ribosome to retreat from the decoding format. Next comes the Amber Stop, which amplifies into Hexagram 56 ䷷ *Traveling Carefully*.

This cautionary message signals the ribosome much like an amber caution light does for traffic. The last traffic codon, UGA, is the surest, most final stop, and it amplifies into the complete stop of Hexagram 12 ䷋ *Standstill* which gives the effect of a red traffic light. Thus we see that all three stop codons exhibit an equivalent message in their hexagram meaning.

Next let's turn to the Met-Start codon and its Hexagram 41 ䷨. Translations render this hexagram as *Decrease* or *Sacrifice* or *Reduction* or *Initial Loss for Later Gain*. More than most hexagrams, its dynamic is hard to grasp for the modern mind—as Legge points out when he says, "The interpretation of this hexagram is encompassed with great difficulties."

But I have kept careful records of its appearance in my personal events for a number of years, and experience has shown that it signals to decrease extraneous, unproductive motion and redirect my energy to begin something of value, but by starting small. This dynamic is aptly shown by the name of *Starting Small* or *Conception*. If you read Hexagram 41 for the underlying concept (much as Legge suggests in his Introduction), you will see that it indicates that if you make a small sincere beginning, it will bring success; but to start it, make a small offering of two containers of seed.

Symbolically, I liken these two offerings of grain or seed to the parental sperm and egg. Life starts with two small contributions that plant a new life. It involves a temporary sexual energy investment/depletion for later gain. The philosophy of Hexagram 41 ䷨ *Starting Small* thus corresponds to the start-codon of AUG, the go-traffic codon that starts a ribosome at its tasking. The last section of this book shows an example of how one might work with the archaic, noble style in this hexagram to grasp its essential meaning for today.

When a hexagram meaning is difficult for the modern mind to grasp—as with Hexagram 7 or 9 or 41 or 47, for example—I find that sometimes the commentary of various scholars through the ages has become padding at the price of accuracy. The sharp shock of recognition is lost. Commentators overwrite the original succinct text and take it in a different direction, related to be sure, for analogs always relate, but sometimes not quite ringing the old truth. Wilhelm, for example, in Hexagram 41 appends commentaries from the *Ten Wings* that are sonorous but occasionally spin out an image beyond its tensile strength. To me this suggests the danger that was in old China's dominantly analog society forever ringing changes on its old habits without using an occasional sharp edge of logic to hew through the clotted past and get things back on course.

Legge's terse text in Hexagram 41 sticks closer to the original judgment of King Wen—although it's not very poetic. "In (what is denoted by) *Sun* [Hexagram 41], if there be sincerity (in him who employs it), there will be great good fortune:—freedom from error; firmness and correctness that can be maintained; and advantage in every movement that shall be made. In what shall this (sincerity in the exercise of Sun) be employed? (Even) in sacrifice two baskets of grain, (though there be nothing else), may be presented."

By the way, Legge is almost unique in putting parentheses around each word or phrase that he himself adds between the Chinese characters to tie them together. Most translators add liaison words without signalling which parts are their own.

Let's correlate some other exploded RNA messages with the qualitative effects of their amino acids. Tryptophan RNA hexagram is No. 35 ䷢ *Easy Progress*, formed by only one codon, UGG. It relaxes and eases the body and mind. Tryptophan is the ingredient in turkey that puts us to sleep after a holiday meal. In Europe I have suggested moderate doses of Tryptophan to nonpregnant clients as a natural relaxant and antidepressant to aid in sleeping without medication. One takes it on an empty stomach near bedtime instead of using a prescription drug. It is also a natural antidote for migraines. It produces serotonin and niacin in the body, resulting in calmness. At this writing, Tryptophan cannot, however, be bought over the counter in the United States because a contaminated batch made in Japan some years ago was responsible for the deaths of 28 people, so the FDA took it off the safe list. But contaminated eggs and milk have killed many more people, yet that does not make eggs and milk inherently lethal. Nor is Tryptophan inherently dangerous. Used properly, it is beneficial. I hope it will return to the market for safe use.

Since wobble occurs, sometimes an amino acid can be made by several different codons, and therefore its qualitative action is more mixed. For example, Lysine is coded by AAA and also by AAG. The codon AAA becomes Hexagram 1 ䷀ *Creation*. Codon AAG becomes Hexagram 14 ䷍ *Possession in Great Measure*. To block the debilitating effects of herpes, shingles, and cold sores—all versions of the same virus—try Lysine. It works quite well. I have suggested that clients take 500 milligrams to 1 gram a day for maintenance and up to 4 grams spread throughout the day during an outbreak. KAL Pharmaceuticals has pointed out, "The theory says that extra lysine forces the virus to lie dormant in nerve cells. Benefits are that herpes outbreaks are more mild when they do occur, and, if regular daily supplementation of 1 gram or so of lysine is continued, the outbreaks will occur very infrequently—perhaps never." It is important to emphasize that Lysine does not kill the virus, but merely keeps it inactive and suppresses the symptoms. When a sufferer takes Lysine, the results truly do seem like a creative boon and possession in great measure.

Lysine has another trait that I find amusingly apt for *Possession in Great Measure*. Lysine stimulates the growth knobs at the ends of the bones and prevents dwarfism. Having enough Lysine lengthens the skeleton properly. Truly a possession in great measure.

Another amino acid signaled by two codons is Aspartic Acid. A supplement of this is taken mainly for two things—to eliminate a surplus of ammonia from the body when it becomes toxic and slows down the system, and for athletes to improve their stamina. GAC correlates with Hexagram 28 ䷛ translated as *Great Excess* or *Overbalancing Weight* or *Critical Balance*. This symbolizes a heavy burden causing an imbalance that needs redressing. It may be likened to the surplus ammonia burdening the body and putting it out of balance.

One night I went to bed particularly tired and said to myself, "Oh, I need some help!" That night a short dream showed a vague picture of "too much ammonia," and a voice said "Get rid of the ammonia." Groggily I woke up: "What a stupid dream! Doesn't mean anything." But fully awake, I realized it must, so that very day I went to a Swiss pharmacy to buy some Aspartic Acid. The pharmacist had never had a call for this before. But I felt better after taking it.

Aspartic Acid's other codon of GAU correlates with Hexagram 32 ䷟ which is translated variously as *Duration, Continuing, Enduring, Constancy, Continuity*. Within such names, you can see a concept that tallies with Aspartic Acid's ability to improve athletic stamina and endurance. Indeed, anyone can benefit from it.

But adding Aspartic Acid to the amino acid Phenylalanine in a special non-digestible form makes the artificial sweetener Aspartame. I do not recommend this because it triggers PKU syndrome in infants who have a pair of genes with a metabolic variation. Some adults (including me) also feel adversely affected by Aspartame, and experience it causing memory failure and various health problems. A national organization called Aspartame Victims and Their Friends (AVTF) is devoted to changing the food laws regarding Aspartame. Toward all such body-flummoxing fake foods, I take this policy: "Why trick my body into eating what it can't digest? This fake stuff does not just go right through—it taxes my body and causes trouble in the long term."

Phenylalanine combines easily with various other substances by playing a secondary role to more dramatic things like adrenaline and norepinephrine and dopamine. Its codon UUU correlates with the all-yin traits of Hexagram 2 ䷁ *Receptive,* which is being receptive and doing work while staying in the background.

Phenylalanine also helps your blood circulate better. This tallies with its other codon of UUC, which matches Hexagram 8 ䷇ *Holding Together,* whose trigrams of Water over Earth convey the image of rivers binding the earth together in their bright tracery of fluid. Yes, Phenylalanine has qualitative traits analogous to this hexagram.

Histidine is coded by CAC and CAU. Now CAC will decode into Hexagram 49 ䷰ *Revolution,* in which Fire sits under Lake, burning and boiling so that it seethes with revolution. A Histidine imbalance brings trouble in the brain's inhibitory transmitter. For instance, too much Histidine is found in the brains of schizophrenics. And yet a Histidine deficiency will deplete the natural alpha rhythms in the brain and allow the excitory beta waves to dominate, resulting in anger and tension.

Psychologists know that arthritics often show a pattern of inner rigidity and repressed rage. Their blood boils with anger. Studies suggest that the tension of holding in rage can somaticize into stiffness and pain at a physical level. Dr. Donald Gerber in Downstate Medical Center in New York has found that the blood of arthritics contains only about one-fourth as much Histidine as the blood of healthy people, so he tried giving large doses of Histidine to arthritics. He found that it "seemed to benefit these folks greatly. Some of them showed improvement with only one gram of Histidine daily, others needed as much as six."

Histidine's codon CAU equates to Hexagram 55 ䷶ *Full Abundance,* whose trigrams are Fire and Thunder. This dynamic shows a fullness of energy like the sun at its zenith. By association, perhaps it helps to avoid the stiffness of arthritis is by cultivating a flexible, generous psyche that is also able to express anger instead of storing it up. Histidine soothes the stomach, relieves heartburn, nausea, and even heals stomach ulcers. *Full Abundance* also recalls that Histidine can't be manufactured by the body, unlike most other amino acids. It must be supplied by a full abundance of food.

Histidine not only strengthens your mind, but it also strengthens your immune system by regulating antibodies so they can fight allergies, viruses, and toxins. And Histidine can even can bring sexual abundance! Women who cannot achieve orgasm have been given more Histidine to great effect, while paradoxically, men who have premature orgasm show too much Histidine. This can be regulated by taking methionine and calcium.

Correlating the philosophical qualities of hexagrams with amino acids may seem farfetched and illogical. You're right, it is. Qualities are not logical, but rather, analog, connected by associations that hold things together in networking resonance. The Chinese used to phrase this principle as the question, "What likes to go together?"

Other amino acids can be correlated with their RNA hexagrams. Some are coded for by so many codons, though, that it might not be very fruitful to cross-correlate unless one could somehow ensure that a particular quality is actually triggered by exactly that codon. The problem is the wobble nature of the last base, so that some amino acids are coded for by four or even six codons—for example, CCU, CCC, CCA, and CCG all code for Proline. Biology calls this wobbly coding redundant and therefore degenerate—as though it has no hidden purpose. Labels such as "nonsense codon" and "degenerate coding" suggest to me a blind spot in the scientific mindset which declares that since the utility of something is not yet apparent or understood, it does not exist.

Normally biologists view most of a gene's coding as degenerate and redundant. They also think all the various codons for an amino acid form the very same substance, but I wonder about that. Perhaps it just appears so at our present level of knowledge. I also wonder if a specific hexagram property is generalized among all its wobbly coding variables of an amino acid.

For example, among Arginine's six codons, is it actually only the AGG version of Arginine that enhances male potency? AGG is Hexagram 38 ䷥ *Polarity*, of Fire over the Lake creating a polarized tension that can bring a clash of wills, or if channeled properly, a union within polarized attraction. Often it is sexual, as portrayed in *Kiss Me, Kate!* or "Vive la difference!" It fosters both the battle of the sexes and also intercourse. It recalls the interplay of masculine and feminine that we saw earlier as the two interconnected triangles that together create one star.

Here's another odd point about Arginine. You recall that Lysine inhibits an outbreak of herpes—well, Arginine actually encourages one. What an opposing clash here! To keep a high ratio of Lysine-to-Arginine in your body, avoid too much high-Arginine foods such as chocolate, coconut, sardines, nuts, and carob. But don't try to eliminate it completely! Arginine builds muscle, burns excess fat, heals wounds, increases sperm count and motility. If you want the benefits of Arginine without the Lysine-inhibitor trouble, you can take the supplement version called Ornithine.

I won't attempt a complete cross-referencing between each RNA message hexagram and its amino acid properties. However, enough amino acids and traffic codons have been evaluated qualitatively, I feel, to suggest that hexagram meanings and amino acid properties do dovetail in a manner far beyond chance.

We can even begin to get an inkling of how Chinese medicine and acupuncture find their roots in holistic process. It is not very accessible to the Western mind trained on linear logic. But it certainly is "rational"—literally so. It uses analog ratio. It exists in the holistic network that promotes qualitative process rather than quantitative product. Chinese traditional medicine, like Chinese philosophy, treats nature as a hologram and works to harmonize the ever-shifting balance among the parts of a constantly evolving universe. It is a very different way of conceptualizing reality. It offers the West something we need to expand the limits of our linear, logical reality.

Learning the I Ching expands the mind's boundaries. How could one possibly code anything more universally, simply, and economically than with yang and yin? If you were to inscribe this code for another culture or another world, what more efficient symbolism could you use? It writes out the whole system using only two kinds of lines, •• and ▬. This is a far more universal symbolism than the alphabet letters of A, T, G, and C. Its format implies far more information. Furthermore, it tolerates primitive methods of recording such as scratches on bone or in dirt with a stick. Its sophistication lies in the mind, not in the medium. The code itself meshes linear and analog number into a transcendent third condition.

But wait. Something's still left unresolved. We have an order of hexagrams still unexplained. It is the dynamic, changing order of the Chou I, used for consulting the I Ching. What could this shifting, mutating property refer to in genetic terms?

To make a guess—and why not?—consider its ancient history. Long ago there was not just one book of the hexagrams, but rather, three. The *Official Book of the Chou* dynasty says that in ancient days the Grand Diviner had charge of the rules for three different systems. They were called the *Lien Shan* (the Mountains Meet) from the Hsia dynasty, the *Kuei Tsang* (Return and Conserve) from the Shang dynasty, and finally the *Chou I* (the familiar Book of Changes) from the Chou dynasty. In each system, "the regular or primary lineal figures were 8 [trigrams], which were multiplied in each till they amounted to 64 [hexagrams]."

These three systems are mentioned twice in the old *Official Book of the Chou*, according to James Legge. But we have little record of what those two lost books contained. A few texts from the Han dynasty mention them, claiming that Lien Shan had 80,000 words and Kuei Tsang had 4,300 words. Unfortunately, for all its love of scholarship, China also has a history of destroying books when power shifts in the government have deemed certain manuscripts "dangerous." However, Chinese ability to protect and further scholarship has rated far higher than in most ancient cultures. Nonetheless, the occasional purges brought disaster to learning.

For example, there was a great book burning and beheading of scholars after the arrogant thirteen-year-old ruler of a western state called Chin began conquering the neighboring states "like a silkworm devouring a mulberry leaf" in the words of ancient historian Su-ma Chien. Finally this brash but gifted ruler united them into an empire that would become known as China.

This boy ruler, Chin Shih Huang, became the first emperor of Chin, and putting an end to every kind of dissension was his goal. He standardized laws, writing, coinage, weights and measures, wheel gauges, built a network of roads and canals, developed the Great Wall of China—and killed 450 Confucian scholars whose philosophy he did not like, along with burning many important books. Although this remarkable man accomplished so much in the unification of the new China, the Chin Dynasty itself faded away within four years of his death.

Yet this man left an indelible mark on Chinese history. His brief dynasty even named the country itself . . . perhaps because of his flamboyant and dynamic nature. He was that extravagant emperor whose guardian army of 6,000 life-size pottery men and horses accompanied him into death at the huge park-mausoleum that he constructed during a period of 36 years at Mount Li. The pottery army was excavated to the delight of archeology and art in 1974. I hope that eventual excavation of the emperor's tomb at the northern foot of Mt. Li may even reveal something new about the I Ching, for it is possible that copies still do exist inside, since this "useful" book was spared the emperor's wrath so long ago.

How did the I Ching that we use today escape young Chin Shi Huang's great book burning? According to James Legge, "In the memorial which the premier Li Se addressed to his sovereign, advising that the old books should be consigned to the flames, an exception was made of those which treated of 'medicine, divination and husbandry.' The Yi [Chou I] was held to be a book of divination, and so was preserved."

Nevertheless, we have lost most information about the three old books. Some people say that perhaps they were not really three separate "books"—after all, a book in those days was a much shorter affair than we generally mean by the term nowadays. Maybe they were just three separate hexagram orders. Well, hexagram order really does matter, as we have seen. After all, it is specifically the Old Family order that we have cross-coded with DNA.

Happily, during the 1970s an unfamiliar order of hexagrams did turn up suddenly—written on silk and therefore called the Silk Text. It came from a 168 B.C. tomb that was excavated in Mawang Dui in Hunan Province.

This Silk Text has a whole new look. For one thing, its hexagram sequencing is different. For another, it has no instructions on divination, so perhaps this version wasn't even meant to be used in that way. Third, there are some textual differences in hexagram meanings. Greg Whincup's interesting book *Rediscovering the I Ching* takes this new Silk Text into account.

Silk Text

The Silk Text order springs from four sentences found in its Great Treatise. Liu Dajun of Shandong University in China translates them this way: "Heaven and Earth establish their position. Mountain and Lake interpenetrate their chi. Fire and Water overcome each other. Thunder and Wind influence each other."

Heaven, Earth, Mountain, Lake and so on refer of course to the eight trigrams. They show the order of the bottom trigrams along the top row of the hexagram chart, reading from right to left in typical Chinese fashion. First comes the lower trigram of Heaven; second, Earth as its complementary partner. Therefore this Silk order keys the lower trigram relationship not just within a hexagram, but also *between* hexagrams. Next, Mountain and Lake lie beside each other to "interpenetrate their chi" or in modern terms, interface their energies. Next come Fire and Water, so strong that they overcome each other and switch places. Finally the last two, Thunder and Wind, sit side by side influencing each other. The result for the Silk Text is an intricate yet harmonious mesh of trigrams.

This rather poetic explanation of the Silk Text trigram order is typical of Chinese nature analogy. One finds it everywhere in the ancient writings. Look, for example, at a Tang dynasty (618-906 A.D.) equivalent of the *Kama Sutra* called Tung Hsuan's *Art of the Bedchamber*. In it the author calls the penis by such nature-oriented imagery as the Celestial Stem, the Jade Root, and the Male Peak, while he indicates the vagina as the Precious Flower, the Dragon Jade Gate, and the Golden Gully.

Soon enough, by studying these pairings, it becomes evident that this Silk Text version employs no stable Old Family trigram order, but rather, something that is more akin to the dynamic order of the New Family trigrams.

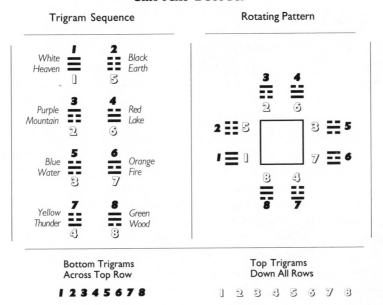

Silk Text Decoder

185

This Silk Text Decoder, modified from the work of Liu Dajun, shows the varying rhythm of the upper and lower trigrams. Gao Heng says, in an article from *Wenshizhe*, that "This type of sequence made it easy in performing divination to find in the text of the Yijing [I Ching] a given hexagram and the hexagram into which it changed, but this is only consistent with the needs of divination and does not have any philosophical significance."

But personally I wonder whether this new Silk order may not be significant indeed. I think it is quite possible that these three "books" show not just three arbitrary orders, but even three different subcodes of the 64 equations called hexagrams. We have already seen that the Old Family order unlocks the RNA codons. I wonder what the other orders might suggest in genetic terms? I don't know. But it's fun to speculate.

What about this third book, the hexagram order that wasn't lost? It is our familiar Chou I, the Book of Changes known nowadays as the I Ching. Its hexagrams are paired two-by-two along the rows. A simple symmetry rule holds each pair in partnership, so that its second hexagram either mimics the first by returning (standing on its head), or it mimics by changing (reversing its line polarity). You can see this happening in the chart on the next page.

Most pairs return—in other words, they are inverted versions of each other . . . 24 pairs, to be exact. These are the 24 freestanding hexagram pairs without Pair Boxes in the chart. But 4 pairs change: they switch line polarity into negative images of each other, and they have simple Pair Boxes around them. Finally, there are 4 versatile pairs, where the second partner can be see in either option: either it returns, i.e., stands on its head, or it becomes the negative image of the other. These last 4 versatile pairs sit in the Split-Pair Boxes.

But this hexagram pairing system is doubly felicitous; it works not only mathematically, but also philosophically. In conceptual paradox, one partner becomes a mirror-twist commentary on the other. For example, Hexagram 49 ䷰ *Revolution* boldly and abruptly overturns the established order, but its return partner Hexagram 50 ䷱ is the slow-cooking *Cauldron* which only gradually changes things through evolving development—it is likened to cooking food that slowly transforms from a hodgepodge of raw ingredients into something digestible and palatable. All the paired Chou I hexagrams contain this entrancing mirror symmetry of trait and form.

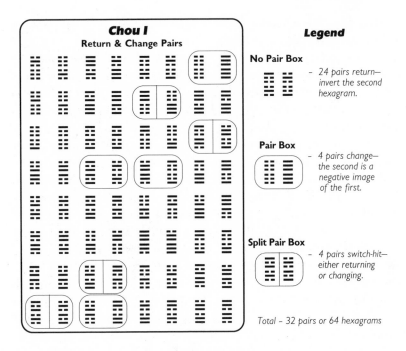

More important than this paired symmetry and meaning, though, is the fact that the Chou I sequence is set out like a storyboard. It begins when boy meets girl—I mean when the creative sky father of Hexagram 1 meets the receptive earth mother of Hexagram 2. Together they generate an offspring who goes through the difficulty of birth in Hexagram 3, an arduous beginning which is, however, supremely successful, for it launches the newborn into Hexagram 4, where the child faces the daunting task of climbing the high mountain of learning about the life it has been born into. Hexagram 5 teaches the necessary patience of waiting as it learns to delay gratification. Hexagram 6 brings the test of personal conflict. Hexagram 7 turns this personal conflict into the strength of a social support system in time of conflict that is submerged in the environment itself, reliable and strong. Hexagram 8 brings the growing child to a more conscious state that gladly holds together with sublime values, a constant heart, and firm effort. Hexagram 9 tests this new-found resolve by accumulating a powerful tension that does not discharge in relief, much like heavy promising clouds that don't let fall their rain in time of drought. Hexagram 10 teaches the child how to tread on the tail of the powerful tiger of dangerous authority and not get bitten in this fecund jungle of life.

And so the story goes on through the entire 64 hexagrams of the Chou I order. Its plot line is more concerned with the inner life of the developing psyche than with external values like wealth or social status. There is a slow growth through the joys and tribulations of adolescence into adult responsibility. Always the inner life is paramount, the inner search of motives and mental images and a humble sensitivity to the patterned way of the Tao that is beyond mere logic.

Toward the end of the sequence, Hexagram 60 accepts physical limitation and works with it. Hexagram 61 honors the vivifying core of truth that sits inside like a chick inside a precious egg. Hexagram 62 is conscientious in small things, not just the larger issues that had gripped the attention earlier. Then in Hexagram 63, *It's All Over,* everything looks like it's all ending, completed, tied up in a tidy little bundle with everything "in its proper place even in particulars," to quote Wilhelm. But then in Hexagram 64, *It's Not All Over,* one discovers that this ending is only the start of something new. Thus, the dynamic never ends.

The ancient Chinese synopsized all this drama into a little verse that describes the story, hexagram by hexagram. Centuries ago, school children knew all the hexagrams by this rhyming verse—much as our children today know a Coke jingle or *Humpty Dumpty.* Verse is marvelously successful in getting right past the logic-choppers of the left brain because it resonates the right brain into coming on-line. The ancients well knew the mnemonic value of verse. It is why a silly advertising slogan lurks in memory. It is why Hitler put his propaganda into song. It is the heart of rap music.

It seems to me possible that, going by what we've already seen, this last book may even be called Book of Changes to significant genetic purpose. Italian and Spanish actually translate the title as the *Book of Mutations.* I suspect that within the yarrow stick manipulations, we perhaps possess an analinear system for describing genetic mutation and its rhythms of change. Nobody yet understands biologically the strangely foresightful nature of genetic mutations. I do not know how to apply the I Ching mutation rules here to genetics. We may even speculate that its algorithm may actually parallel some process whereby purposeful mutation occurs in RNA, which somehow furthers the physical evolution of life. We already do know that the intent of the I Ching is to further the life of the mind. An application to genetics would give a teleological goal of matter to this third book called the mutating order of the ancient Chou.

Just as the genetic code carries messages to build the body, so does the I Ching carry its own code for evolving the psyche. It whispers about a patterning that's written not in the hard flesh of generations but rather, in the intangible flow of mind itself. It codes for thought structure, at the personal level, at the deeper level of all organisms, and even of the living cosmos. It suggests some total mind in the supersystem that presents choices and options, if we can just become aware of them.

Thought structure is more malleable, more able to change than the proteins that make tangible fiber and flesh. This is fortunate, for changing your mind can alter your matter, and the matter of events coalescing around you. The two realms of mind and matter merge somewhere back there behind the veils of unknowing. They unite in co-chaos, and following the Tao leads to this ultimate source.

Some may say that speculating on the lost books is too fanciful a flight. But it was just such idle speculation that led me to wander into co-chaos. It has been a great delight. Conjecture on the two lost books is fun and unprovable, but it is only a preliminary to introducing my real point here, which is not speculation at all.

An efficient code employs multiple-layering to conserve space and energy, and in fact, a truly efficient master code can be applied at various levels. The hexagrams are a master code that can be applied to the gene, yes, but also perhaps to other levels of physical science such as the electron arrangement in atoms, and even to the level of quarks and leptons. I suspect that the genetic code is merely one application of this much larger paradigm of co-chaos where two Period 3 windows interface in analinear equations. The possibility of uniting physics and metaphysics begs for more exploration. A large and heavy topic indeed.

But in the next chapter let's go light and just have some fun with number as artful image. After the weight of our genetic heritage in this chapter, why not take a lesson from the I Ching's reversals? Let's just play around with analinear images for awhile, some patterns for the sheer pleasure they bring. We'll explore some subtle beauties in the next chapter, "Art in Number and Image." We'll experience the co-chaos paradigm not as science, but rather, as art.

WORKSHOP 9

Plan a program that combines discussion and experiences. Select book paragraphs to introduce the following topics—or your own.

Opening ritual.

❧ Ahead of time, ask a member to prepare a brief talk on Chin Shih Huang Ti and his pottery army. (See *National Geographic,* April 1978 or other sources.) Or pick a Chinese spot seen by a member.

❧ In the group, discuss the traffic codons and their messages. What do you think of their genetic and philosophical correlation? Is this possible? Significant? What of the correlation between amino acids and physical symptoms? Do you have a particular body ailment? Have you investigated to see whether an amino acid could help it? How could you explore this safely and thoroughly?

❧ Ahead of time, ask two members to pick a hexagram discussed in this chapter and together prepare a presentation on it using an I Ching book or pp. 263-272. After the presentation, discuss the hexagram and its implications. What does the presentation suggest to you about the ancient Chinese mind? About your own?

❧ Discussion: do your dreams ever give you messages that you are able to utilize in everyday life? Do you think they even *can*? Poll the group. What are the significant differences between the dream world and the waking world? Is there a way to make these two domains mutually supportive? How can you try it and see?

Meditation (to music): Close your eyes and go back to the last dream that you remember having. Even a tiny fragment will do. Hold this dream in your mind and watch it as if it were a movie, study it as if it held a mystery clue. What could it be telling you, showing you? Your unconscious mind knows the answer. Slowly go deeper into yourself and let its message rise up spontaneously within. Ponder this message, even if it seems silly or contrary to your ego's inclination. Seek to apply this dream clue to your real life to draw your conscious and unconscious worlds together in mutual support. Consider making an effort to open to each dream and remember it, understand it, heed it, thank it, apply it.

Closing ritual. Five minutes of feedback. Announcements.

Section 3

Transcending the Twain

The Pleasures of Merely Circulating

The garden flew round with the angel,
The angel flew round with the clouds,
And the clouds flew round and the clouds
     flew round
And the clouds flew round with the clouds.

Is there any secret in skulls,
The cattle skulls in the woods?
Do the drummers in black hoods
Rumble anything out of their drums?

Mrs. Anderson's Swedish baby
Might well have been German or Spanish,
Yet that things go round and again go round
Has rather a classical sound.

Wallace Stevens

Art in Number and Image

I call this image "Bringing in the Sheaves." I also call it the 64 pairs of Period 3 windows in co-chaos, rendered on three scales.

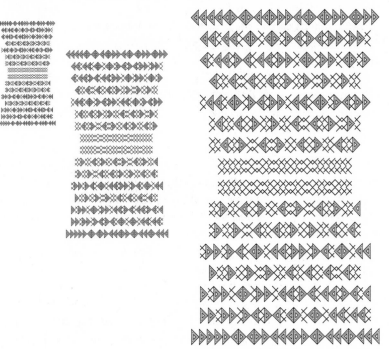

Bringing in the Sheaves

How did we leap from hexagrams to sheaves of grain? Perhaps if you look at the image as something like needlepoint, you'll guess. A needlepoint design can be blocked and filled in according to a code or legend. Here we are using the I Code as our master design.

Although artists don't usually think about it this way, the basis of design is number. Number is the basis of color, too. And of music. In fact, underlying all the arts and even all artistic sensibility is number. Analog number. Analogs get together to discuss the relationships between proportions and fit; they provide each component's relative impact in the whole. These whispering resonances carry on a dialog that makes things seem graceful or awkward, elegant or sloppy.

But an artist generally doesn't view number in this way. In the artistic sensibility, number seems cold and mechanical and soulless. Yet number can be passionate and organic . . . if it is analog number, which values relational quality over finite quantity. Such number prefers proportional shifts in nuance over summary statements. Such number caters to taste more than to fact. It is subjective rather than objective. It glories in impression and response, not in analysis and conclusion.

But even art can be more or less linear, less or more analog. When I lived in Turkey many years ago, I found it saturated with a music that I couldn't even hear very well, much less appreciate. Wailing fife, heavy drum. It slowly began to intrigue and enfold me in its circling textures and finally to delight me in a wholly new way. So it came as no surprise to discover that these swirling repetitions of endless relationship are an Anatolian heritage that was eventually brought westward to Greece in the cult of Dionysius. Dionysius, the dark unconscious opposite to cool linear Apollo. Dionysius, with his brow encircled in a wreath of iterating grape leaves, inebriated with body and soul, lost to reason and logic, circled by dancing maenads. Yes, Elvis lives . . . as *deja vu*. A Dionysian music stirs and captures its listeners and sometimes even drives them to mad abandon with its driving beat and note clusters that cycle again and again in self-similar, resonant, not quite repeating effects on the brain. Such music does not really end. It has no final climax or resolution. It just becomes lost to the ear as its pattern sinks below conscious hearing.

Such music has a cumulative hypnotic power far different from traditional Western music, which is much more linear. With my ear trained to the heroic Western mode of conflict, crisis and resolution to a goal, I didn't know how to surrender into this Anatolian swirl and let its endless process enfold me, let its cycling iterations carry me into a network of emotions.

Jonathan Kramer has pointed out that Western music is linear to a substantial extent: "Tonality embodies a set of hierarchic relationships between tones, supported by durations, dynamics, timbres, and so on. The tonic is endowed with ultimate stability. All tonal relationships conspire toward one goal—the return of the tonic, finally victorious and no longer challenged by other keys. Thus tonal motion is always goal directed." This climax-oriented music recalls the frantic car chase in a Hollywood movie. Or the orchestrated greed of TV game shows. Or pep rallies and sales quotas. It's all a glorification of that line-driving thrust, of yang heroics in pursuit of its goal.

But let's quit going for the goal and just noodle around in artful analog circles for awhile. Let's relax and loll along the banks of analog number, picking out small designs and watching fractal castles in the air. It's amazing how pleasurable this can be. We'll discover co-chaos in a new way, not as a code to be broken, but as a design to be savored. Remember, these are offered for the art of it, for fun.

Look at how subtle and harmonious are the patterns that hide in the following Old Family Order of codons! Here is no stodgy textbook format, but a smoothly folding swirl of relationships. From each corner a motif is introduced that weaves toward the center where the four-part theme is confirmed in a central arabesque of complementarity. (The traffic codons are underlined.)

Codons in Old Family Order

CCC CCU CUC CUU UCC UCU UUC UUU
CCA CCG CUA CUG UCA UCG UUA UUG
CAC CAU CGC CGU UAC UAU UGC UGU
CAA CAG CGA CGG <u>UAA</u> <u>UAG</u> <u>UGA</u> UGG
ACC ACU AUC AUU GCC GCU GUC GUU
ACA ACG AUA <u>AUG</u> GCA GCG GUA GUG
AAC AAU AGC AGU GAC GAU GGC GGU
AAA AAG AGA AGG GAA GAG GGA GGG

Notice how the pattern of bases swirls toward the center. But maybe it's easier to sense if our codons are indeed arabesques. Following are the Old Family codons rendered in a different symbolism.

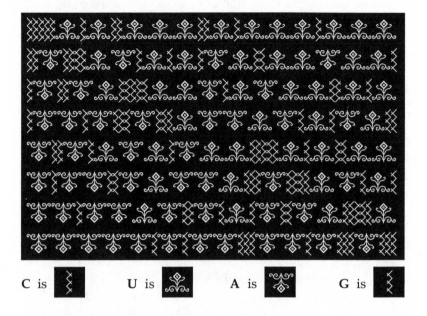

C is [symbol]   U is [symbol]   A is [symbol]   G is [symbol]

One can see the proportional relationships in the design better when the three-letter "words" don't get in the way. Now the code is buried in the texture, as it is in the feathers on a bird or the ring of spectral light on a oil-filmed puddle. Symmetry is easier to see at a glance in this format. It emphasizes the analog rhythms in the weave of its design and invites all sorts of associations. What does this dark rectangle look like to you? To me it's rather like a scarf. Does it suggest a title? Personally, as soon as I saw it, into my head popped a tune: "Heard It Through the Grapevine." Later on I also recalled the Silly Centimeter vine that grows chaotically yet predictably. Thus although the underlying mathematical structure is still intact, this analog approach to the codons will resonate the holistic right brain, not the unitizing left brain. We're going arty, not nerdy.

But let's switch gears again and turn back for a moment to view the linear structure. In the Old Family codon chart, there is a binary sequencing all around its outer border. It is easiest to see in a number format. The left edge of the chart on the next page counts in binary code from 0 through 7. But actually, all the edges do that, just using different symbol designations for the binary symbols. All the edges employ the same distribution pattern.

Number Distribution Chart

C is 0	000	002	020	022	200	202	220	222
A is 1	001	003	021	023	201	203	221	223
U is 2	010	012	030	032	210	212	230	232
G is 3	011	013	031	033	211	213	231	233
	100	102	120	122	300	302	320	322
	101	103	121	123	301	303	321	323
	110	112	130	132	310	312	330	332
	111	113	131	133	311	313	331	333

This is the codon chart rendered into binary and analog number. It actually reveals the binary code of Leibniz presented four different ways, but it's arranged so that the four different binary formats are also interrelated in analog ratios. Notice that each of the four corners contains the ultimate expression of one of the four symbols. Gradually each symbol feeds in across the pattern as a smooth sequencing. This happens both from edge to edge along the four sides and also diagonally across the weave. Each inward proceeding layer furthers the interwoven process of relationship.

So this chart is not just the ordinary binary progression that is visible on its outer edges. Instead it is also an analog pattern that is woven throughout, using the four genetic code bases, here called 0, 1, 2, and 3, as the weaving material. This same dynamic also occurs of course in the polarity system of the I Ching. This woven order is in fact one of the best succinct proofs that the I Ching and the genetic code do use the same mathematical paradigm, and that this structure is both binary and analog in its structure.

Many people comprehend such things better, though, with their artistic intuition rather than logical thinking. So let's make some designs in the genetic code. Its rhythms will appear in various computer graphics that could be even more nuanced by adding color or sound effects. The mind continually generates just such designs in arts and crafts. This next pattern reminds me of a rug.

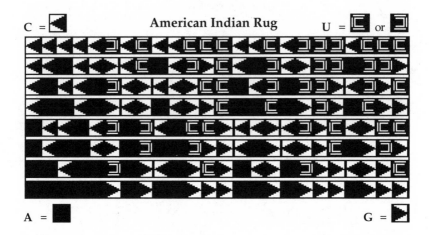

Interestingly, my old printer balked at reducing the concentric square image that makes up the U symbol. So it iterated this symbol in two partial versions, as a sidewise U facing either left or right. It of course reminds me that this molecule also has two versions, either as T or U, depending on whether it is in DNA or RNA. In this "rug," the squared-off quality tends to emphasize the blocky continuity of the design.

If this basic design were repeated four times and then the blocks were "sewn together" in a mirror-image way, our rug would match symmetrically at all corners and in the middle. This is how folk rugs are often made, in fact. If it were actually done, personally I'd prefer to have my rug blocks matched together at the lower left hand corner, in order to put the darkest values in the center. But that's my own artistic taste that tends to go toward the dark heart of things and turn them into mandalas of order. How about your preference? Do you see any trend behind it?

Next is a lighter, looser-looking design, not quite so squared-off, but still with its components emphasizing the blocky effect. Two mirror images symbolize C and G in this design, but it has two more divergent images for U and A, which tends to loosen up this pattern for a more spontaneous over-all effect.

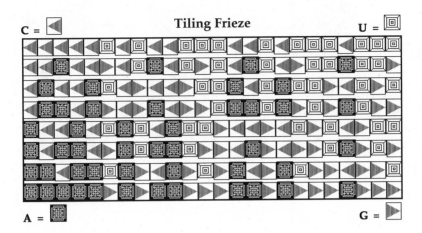

It's amusing to me that this Tiling Frieze is really the genetic code. Which do you prefer aesthetically, this design or the previous tighter, bolder rug on the previous page? Actually it's exactly the same computer image but just rendered at two different resolutions. Notice that in this second version above, my new laser printer didn't squeeze the U symbol into two variant forms. So much of a pattern's effect will depend on the resolution at which we see the details.

This is analogous to observing the passing events in your own life, where occasionally if your vision gets enough loft and perspective, on a clear day you really do even feel like you can see forever . . . in the upbeat lyrical analogy of popular music.

Here comes a yet more fluid version of the genetic code. Its curves use spacing as much as the symbols themselves for encoding. Notice that the effect becomes more dynamic than in the previous two blocky designs. It is the variable spacing that gives this design its sensation of movement. We often call something "fluid" if it has a sense of motion in the static symbols themselves. The motion is implied by seeing the proportions in dynamic relationship.

First, the coding chart. The basic symbol is only one chevron, but it can be flipped to either side, or it can be spaced more forward or backward. So you'll notice that it is position that gives the real code.

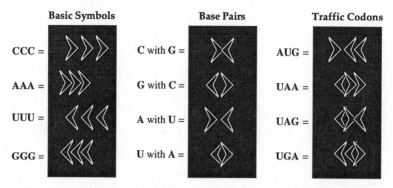

One basic symbol repeats with variation. Position is the key to showing the identity.

The flying wedge is the only symbol used here. Position alone is enough to differentiate it into the two branches of the four bases. The pyrimidines C and U are spaced at the outer limits, while the purines A and G are spaced at the inner limits. The resulting pattern will vary according to which base sits next to which along the row. Their spaces overlap sometimes and isolate each other sometimes. Thus position is the differential encoder here . . . as also happens with the bond-bigrams in a hexagram.

Notice the four traffic codons on the right. These clusters of three-somes all regulate the flow of the protein-making process. To me they have a lyrical quality. In fact, the flow of this design has an almost hieroglyphic effect, rather like telling a story. When I made it, somehow it reminded me of a symphony of classical music where at first people are sitting and listening with their hands folded in graceful attitudes of pleasure and then at the end, they are clapping.

And as synchronicity would have it, after I sat at the computer one gray Sunday afternoon making this musically flowing design, I decided to take a break. During a stroll in Zurich, I realized I'd left my purse at home. Then I noticed an announcement for a free concert for harpsichord and flute using period instruments. In the light snow I hurried on over to the nearby Catholic church, and I was just on time. As the music of Farnaby, Vivaldi, Bach and Bartók filled the air, I imagined fractal patterns such as these cavorting in the high space within the stone cathedral walls.

A Visual Symphony

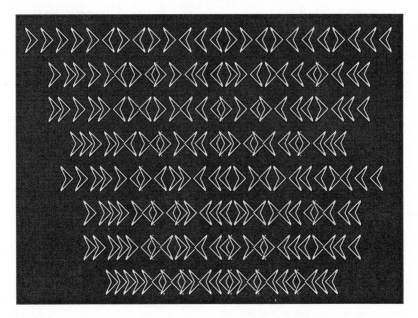

Fractals have natural grace and flow. For so long, the West has lived split between the cool, logical rationality of the classicist and the untamed sensibilities of the romantic rebel. Our culture eventually went so logical and linear that it de-natured the arts into cubism, knife-edged architecture into "living machines," deconstructed poetry into reading the telephone book aloud, chopped dance and music into isolation and discontinuities of raw noise.

But finally we are rebirthing our culture into a new age where analinear vision is able to embrace both sides. It signals an updating of our old standards of philosophy and aesthetics.

All these previous images have been made from codons. Let's see what happens to artful design when we take up the DNA swatch. This pattern will use the paired triplets of the double spiral parents. Through the oppositional looking glass, we discover the result is "As above, so below." Following is a number version of the upper and lower trigrams in Old Family order . . . or in the West's scientific lingo, it's the RNA codons and mirror codons.

The I Code Number Table	Sheaf of Genetic Code

Upper Half

Upper Trigrams–DNA Mirror Codons

A is 1	333	331	313	311	133	131	113	111
C is 2	334	332	314	312	134	132	114	112
G is 3	343	341	323	321	143	141	123	121
U is 4	344	342	324	322	144	142	124	122
	433	431	413	411	233	231	213	211
	434	432	414	412	234	232	214	212
	443	441	423	421	243	241	223	221
	444	442	424	422	244	242	224	222

Lower Half

Lower Trigrams–RNA Codons

A is 1	222	224	242	244	422	424	442	444
C is 2	221	223	241	243	421	423	441	443
G is 3	212	214	232	234	412	414	432	434
U is 4	211	213	231	233	411	413	431	433
	122	124	142	144	322	324	342	344
	121	123	141	143	321	323	341	343
	112	114	132	134	312	314	332	334
	111	113	131	133	311	313	331	333

Sheaf key: 1 is ▷   2 is ⋈   3 is ◁   4 is ⋈

This Number Table and the Sheaf of Genetic Code hold the same co-chaos paradigm. They reflect each other but in different symbolism. In order to emphasize the symmetry, all the lower trigrams are grouped together in the Lower Half; likewise with the upper trigrams in the Upper Half. Notice that this symmetry is caused by the polarity reversal between its top and bottom half. Comparatively, the symbols flip from top to bottom and side to side. Thus, to make hexagrams, the top line of the Upper Half will bond with the top line of the Lower Half. The reason of course lies in the line-pairing rules of the I Ching and the pair-bonding rules in the genetic code.

In the image of the three sheaves of genetic code that you saw at the front of this chapter, the symbols have been scaled down by the computer, of course. Thus they show the tiny bobbles that come during image reduction, which creates slight fractal variations everywhere in the basic pattern, and in fact, you can make fractal scaling just by reducing an image on a copy machine. The lenses inside the machine will distort the image in fractal ways in addition to shrinking it smaller each time. You'll wind up with a Julia set as the limiting image.

In the next table, the trigrams have now been paired into all the 64 swatches of the DNA Plan. Each pair relates through the bindu point between its top and bottom triplet. Each pair is a fraction, but with a bindu point to couple them rather than a line to divide them.

DNA Swatches in Old Family Order

C is 2   U is 4
A is 1   G is 3

```
333 331 313 311 133 131 113 111
 •   •   •   •   •   •   •   •
222 224 242 244 422 424 442 444
```

upper trigrams
• bindu point
lower trigrams

```
334 332 314 312 134 132 114 112
 •   •   •   •   •   •   •   •
221 223 241 243 421 423 441 443


343 341 323 321 143 141 123 121
 •   •   •   •   •   •   •   •
212 214 232 234 412 414 432 434


344 342 324 322 144 142 124 122
 •   •   •   •   •   •   •   •
211 213 231 233 411 413 431 433


433 431 413 411 233 231 213 211
 •   •   •   •   •   •   •   •
122 124 142 144 322 324 342 344


434 432 414 412 234 232 214 212
 •   •   •   •   •   •   •   •
121 123 141 143 321 323 341 343


443 441 423 421 243 241 223 221
 •   •   •   •   •   •   •   •
112 114 132 134 312 314 332 334


444 442 424 422 244 242 224 222
 •   •   •   •   •   •   •   •
111 113 131 133 311 313 331 333
```

Each sixsome of number here is a ratio where all of its parts relate to each other within and across the bindu point. You can look at it as two vertical and horizontal Period 3 windows in counterposed relationship. Take for example that lowest left ratio on the chart.

444
•
111

This configuration whispers of analog ratios, but it tells another story too, that of plain old additive sums. Let's add up a total for this odd hexagram. Across the bottom it's 1 + 1 + 1 is 3. Across the top, it's 4 + 4 + 4 is 12. Okay, add 3 and 12. The total is 15. Now let's look at the totals across the whole row to see how each ratio adds up.

444 442 424 422 244 242 224 222
 •   •   •   •   •   •   •   •
111 113 131 133 311 313 331 333

Hey, each equation across the row adds up to 15! Well okay, let's examine the whole chart. Wow! They all add up to 15! Each equation. Talk about a magic square of fifteen! Equation means equal, and this is it! This chart is 64 different ways of saying 15. All these sums present linear number when we tally up our linear score card.

Now watch what happens when we consider each individual number partnered across the bindu point. In the first bond-pair, 1 + 4 = 5. Likewise with the second. And also with the third. In fact, across the chart, when we add up each pair bonded across the bindu point, each time the total is 5. All over the chart! Somehow there's a whole lot of proportional relationship going on, we begin to see, in what appeared just an additive summing. Number is donning its analog hat again. Tricky, this shifting emphasis in number, isn't it?

Why is the 15 so important? Let's see, there are 8 windows in the analog tree of three bifurcation levels, and then 8 windows in a complementary binary system. It makes 16 in all, unless . . . unless we swing round at the 8th window and let it do pivotal duty and head back toward the start, reaching the origin on the 15th step. Wow, 15 again. It's the same maneuver that the ancient Chinese did in their secret ritual of dancing the "magic square" mandala of the Lo Shu using the "steps of Yu"—named for the emperor of 2200 B.C. who was said to have discovered this secret. Behind this swing point is the reason why both linear and analog number are needed to manage the polarities in this supersystem. Such pivot points are ubiquitous in the many number mandalas of the Golden Mean.

The DNA double helix itself is based on the Golden Mean. This Golden Mean is a ratio using only A and B. It is probably the most famous aesthetic rule, yet it's quite simple, and perhaps so pleasing because of that very simple harmony. To get the proportions of a Golden Mean, you divide a line so that its shorter part is to the longer part as the longer part is to the whole. In other words, split a line into a shorter piece A and a longer piece B in such a way that A relates to B as B does to A + B together.

The Golden Mean

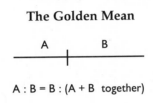

A : B = B : (A + B together)

Notice how this swing term B mediates between the extremes of little and big. The B is the Golden Mean, the common ground, the middle way, the shared note that resonates the whole relationship. It is what makes the proportions harmoniously pleasing.

The Greeks thought the Golden Mean was divine. Literally. They considered its ratio to be the most beautiful fit in nature—physically, aesthetically, even morally. So they took it from nature and put it into their art and philosophy. It appears in their painting, architecture, sculpture and music. It gave the majestic proportions that we now call classical.

Many others have also loved the Golden Mean. Kepler called it "the divine proportion." Descartes charted the way it constructs the smoothly opening spiral of the chambered nautilus. Jakob Bernoulli described this same spiral in logarithms and found it so heavenly that he even wanted it engraved on his tombstone.

This ratio is found everywhere—for example in the center of a sunflower where the seed pad reveals a fifteen-point mandala. The apple blossom takes this same ratio and turns it into the five-point star that is found at the cross-section of an apple. The wild rose shows it in five opened petals. The spider constructs this same soft star as a pentagon by spinning a segmented plane of webbing that spirals ever outward.

Spider Web   **Apple Blossom**   **Sunflower**   **DNA Helix**

Mandalas of the Golden Mean

Here are some natural mandalas of the Golden Mean. The seed pad of a sunflower is basically flat. But the double helix of DNA moves it outward as a third-dimensional thrust, spinning a spiral that can be seen in this molecular rose window that the DNA helix forms when viewed from above. Its symmetry is as smooth and regular and "classical" as any column designed by the Greeks.

But what turns this DNA helix into a such a powerful fractal of patterned chaos is the freight of ever-variable sequencing in its bases. Here again, as with all archetypes, the form endures, but its contents is ever-new, eternally changing.

In fact, as I study the number chart for DNA Swatches on page 203, I see all sorts of linear and analog features playing across the pattern. It is endless, opening again and again to deepening perspectives in a huge fractal series of windows. For example, binary number counts on the top and bottom edges, but decimal number counts each bond-pair adding up to 5 and each pair of bonded triplets adding up to 15. There's also a rhythmic analog sequencing on the left and right edges. Wherever I look, linear and analog number are playing out their complementary dance.

It takes some perusal to discover just how very rhythmic is this number distribution. Its makeup is so diverse and complex that I can't reach the bottom of it. In fact, I'm not even going to show you all that I've found so far; but never mind, you can follow its labyrinthine intricacies as far as you care to go. Here is the very essence of tapping into fractal order . . . however far you take it, windows will open up and reveal new aspects to you.

To think that these fractals occur in the very ancient, inscrutable symbols of the I Ching. A physics student told me that in a book store one day, he saw an I Ching book on display. It showed the hexagrams in Old Family order on the front cover. He stopped and stared at this design for a long time, falling right into it, and he almost bought the book just for its cover, since he knew nothing and cared nothing about the I Ching. The subtle impact on him was that profound. Its obscure mathematical regularity attracted him, those polar permutations that were buried within the design.

So let's just view this Old Family Order now as a design . . . and perhaps we'll even enhance the effect by flipping it sideways to turn off the left brain's conscious tally—turning the yin and yang lines into ripples, perhaps as shadow on a rough brick wall as we walk by. How very enigmatic and economical is this information!

Old Family Wall Texture

American Indian Bindu Point Rug

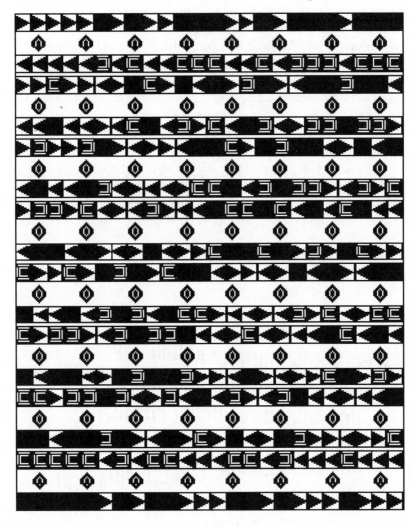

C = ◄ U = ⊏ or ⊐ A = ■ G = ◄ bindu point = ◆

Here is another bold American Indian rug pattern. This design is similar to that earlier American Indian rug made of codons. But now the weave has a new symmetry because both the lower and upper trigrams of the 64 swatch hexagrams are represented in it, and furthermore, they are bonded together by bindu points.

It tickles me that one can read this design as either DNA swatches or the Old Family hexagrams. Each states an equation but in image, not cipher. For example the swatch-hexagram of—

—is found by counting three bindu points up and four over from the bottom left corner. Its lower trigram of or Fire gives AUG, the Met-Start codon. It could later be exploded into the RNA hexagram of *Starting Small*, Number 41. By the way, I didn't plan this rug design to deliver an arrow right here for the start codon. It just happened that way through nature's special penchant for making analog functions that tie the world together in synchronicity.

Synchronicity tries to get your attention with messages from the unconscious. The psyche's analogs have a truth-seeking function in their continual search for comparison and contrast. Through it, the psyche does constant reality-testing. Things finally just will not fit smoothly if reality doesn't allow it. There is no fudge factor here, since psyche spans both the conscious and unconscious realms. This is why sham finally doesn't work in the heart. It may for awhile in the head, if the head is split away from the heart's deep knowing.

Let your head and heart go hand in hand . . . a mixed metaphor by logical standards, but intuition tells you what I mean. Paradox reveals a truth deeper than words. Connecting the heart and head will balance your stance in reality.

Now here's another design that is much more impressionistic. Since it's so wavy and also mathematical, I call it the Bolyai Fabric, named for Johann Bolyai who developed a non-Euclidean geometry in 1823 at the age of 22.

Bolyai Fabric

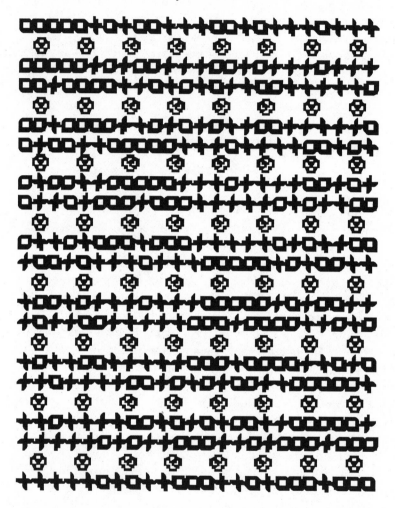

C = ⬛ U = ✚ A = ✚ G = ⬛ bindu point = ✺

There is only one spot in this design where the white space makes a complete eight-pointed white star inside the black. It is found three rows of bindu "roses" down and four to the right inside this design:

And again, although I didn't specifically plan the symbols to create this marker, this star denotes the upper left edge of the three stop codons. Their swatch hexagrams follow the star along the row of—

Their lower trigrams below the three bindu "roses" hold the codons of UAA-UAG-UGA, or written in the script of the Bolyai Fabric—

The serendipitous white space that makes the star suggests that what isn't helps create what is. Tiny synchronicities such as this are always happening as a natural part of the analog realm of number, busy at its subliminal commentary and cueing. Everything networks to resonate in clusters of meaning. We perceive much of the world's pattern through our feeling-tone reaction to it. We know what is out there because of how we feel about it, not vice versa. We are chiming to the resonances hidden in the weave.

This is the root of synchronicity—or "meaningful chance," in Jung's phrase. It suggests rather than declares, it beckons rather than commands. And from these clusters of meaning resonating around a central truth come the group values that we call the ethical and religious and legal norms. Then linear logic usually takes over and tries to spell out and quantify what has already been subtext in the culture. We eventually even tend to go legal and dogmatic with the rules, and try to standardize and organize each other in society.

Yet analogs always run farther, faster, wider, deeper than mere human understanding and rules and society can follow. It outstrips them in its search for truths beyond the codified norm.

Double Weave World

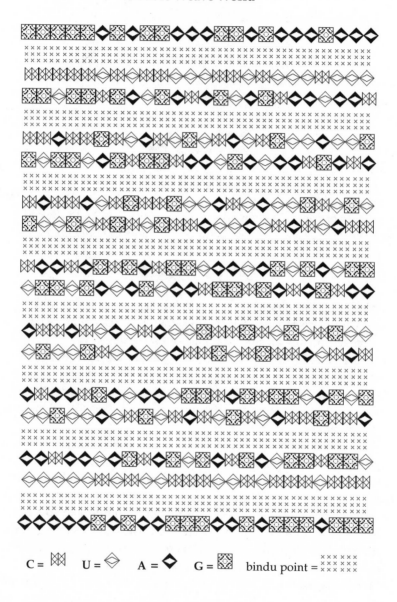

C = $\boxtimes$ U = $\diamondsuit$ A = $\blacklozenge$ G = $\boxtimes$ bindu point = $\begin{matrix} \times\times\times\times\times\times \\ \times\times\times\times\times\times \\ \times\times\times\times\times\times \end{matrix}$

This design to me suggests the interweave of analog number. I suppose it's because the bindu points in it create such a mesh. It's no wonder to me that Indonesians traditionally believe that the universe is formed by the warp and woof of god's thread busily weaving a cosmic design. Many traditional cultures use the carding, spinning, and weaving processes of fabric-making as an analogy for universal creation.

And so does the psyche untutored in all this anthropology, for it carries the collective wisdom in its dreaming mythology. Dreams sometimes feature a great fabric such as a net spread across the sky to describe the universal soul in an image deeper than words. This net of gossamer is a silent symbol for our collective connection, our folk tapestry made of woven souls.

But we also have our own smaller soul fabrics to complete, and we are busily weaving them at a personal level, making the inserts of lace and tangle and rickrack that will blend into the larger whole. Once in a dream I saw some famous people making wind sculptures out of filmy fabric in individual designs that held their shapes yet blew in the wind. They looked something like Tibetan prayer flags that are moved by invisible spirit. And I knew the dream was about some well-known people busy at their soul-making. Analogs mostly billow at the unconscious edges of life, signaling their subliminal messages from the background fabric of events. We take this fabric for granted as the unconscious support system for that tiny fraction of reality that is apparent to our senses and logic.

Numbers form the great wheeling archetypes of transcendent meaning behind the veil of Maya, whose corporeal illusion we put on when we don our bodies for this brief exercise of sculpting our filmy duration in space through time. We become bindu points of matter coalescing the events around us into certain specific proportions and relationships. For a short while, we evolve our individual souls and refine our interweave with the others to take each person's unique place and meaning in the cosmic fabric.

China Seas

C = ◧   U = ◩   A = ◤   G = ▷   bindu point = ✳

This design I particularly like for its crispness. For some reason I imagine it in blue and white, actually as a bedspread—and I call it China Seas. Why? Your guess is as good as mine. Well, maybe not. I do know my own associations. It's just something about these pennants rippling in their semaphores across the wavy lines. It recalls when I lived in South China and frequently looked out across the islands and harbors at boats passing by.

This particular bindu point opens a gateway to China by recalling the ancient symbol of ⌗ Maybe you call it a pound sign or a tic-tac-toe. But in China it was used to indicate a *well*, which in fact is the name of Hexagram 48, *The Well*. This is the well of endless renewal. In ancient days, the central government divided a parcel of land by cross-hatching it so that eight families could live on the eight plots all around the sides, while the central portion was shared in common. This key ninth portion of land held the communal well, owned by all together, and everyone gathered here for the quenching of their physical and social thirst. So this cross-hatched style of land division and allotment became in effect a social version of the magic square.

And that's how it goes with the networking analog quality of number. Its very nature invites analogy. Images suggest something, allusions pop up, clusters of association converge. Details take on relationship. Emerging patterns, being qualitative and relational, stimulate the imagination to reach for more of the design. It encourages insight, creativity. It is the hidden math that generates notions out of the nowhere into the here.

The preceding computer drawings have shown some fanciful ways to graph the DNA swatch hexagrams. They interweave linear and analog number in ways which can even be coded as these pleasing designs. To me these graphics are an amusingly artistic way to show the proportional relationships of number in the I Code.

But now let's turn our gaze to another pattern, to the exploded RNA Writing hexagrams, where the full-blown message opens up from the codon. And just for fun, let's use a new sort of keying . . . although no doubt there is an infinity of possible ways to explore this patterning.

On the left of this chart is the Trigram Key. It counts in binary code straight down the Old Family order of trigrams. Below it are the bindu points that will join upper and lower trigrams into hexagrams. With these tools, we can now explore the "fractions" or fractals of the 64 RNA message hexagrams to discover some emerging patterns. In the middle is a long narrow Trigram Column. Each number here codes for a specific trigram that joins through its bindu point with another trigram to made a fractal hexagram. Thus the archetypal form of the Trigram Column parallels that of the RNA Message Hexagrams Chart to the right, but with different contents.

Trigram Key	Trigram Column	RNA Message Hexagrams
is 0	20202020 · · · · · · · · 55441100	Proline · Leucine · Serine · Phenylalanine
is 1	31313131 · · · · · · · · 55441100	· · · Leucine
is 2	64646464 · · · · · · · · 55441100	Histidine · Arginine · Tyrosine · Cysteine
is 3	75757575 · · · · · · · · 55441100	Glutamine · · Ochre / Amber STOPS! / Opal / Tryptophan
is 4	20202020 · · · · · · · · 77663322	Threonine · Isoleucine · Alanine · Valine
is 5	31313131 · · · · · · · · 77663322	· Met START! · ·
is 6	64646464 · · · · · · · · 77663322	Asparagine · Serine · Aspartic Acid · Glycine
is 7	75757575 · · · · · · · · 77663322	Lysine · Arginine · Glutamic Acid
Bindu Points · · · · · · · ·		

216

These hexagrams are intricately organized by 2 wearing both its linear and analog hats. Everything is ordered through 2-ness. For instance, notice that each upper trigram row (or *numerator row* (like 20202020 at the very top) has only two numbers alternating across it—and this progression adds or subtracts by 2. But each lower trigram row (or *denominator row,* like 55441100 below the bindu points) has its numbers alternating by pairs.

Notice how these denominator rows, when they are viewed as inward progressing pairs— 55441100 —will add to 5 in the chart's top half, but to 9 in the bottom half— 77663322 —and 9 minus 5 gives a difference of 4, or twice 2. Also, both denominator rows skip their count by a doubled beat in the middle—5544 (skip 33, skip 22) 1100 in the upper half and 7766 (skip 55, skip 44) 1122 in the lower. These two denominator rows of 55441100 and 77663322 also show two units difference in value across the row.

The numerator rows also repeat, but using a different pattern of 2-ness. They drop through four permutations and then, midway down, we suddenly see a recurrence of the top row of 20202020. Talk about an intricate pattern of 2-ness! Throughout this whole chart, we can see the continual repetition of 2-ness but with diversity.

It is easier, though, to get an overview of this structure in a more artistic image such as the following whimsical tree. Generally we've chopped up the tree of life and stacked it into dead cords of discrete linear utility. But it is the cycling analog resonances that make nature hypnotically beautiful. For amusement I have developed several designs showing some versions of this tree of life called RNA message hexagrams using abstract symbols for coding. I've named them all as varieties of trees . . . partly since they sprout from codons. Mighty messages from little codons grow.

Here's the legend to decode the first one, the Classical Tree.

Classical Tree

Black Earth is °°꙰°° Purple Mountain is ▨ Blue Water is ▦

Green Wood is ◈ Yellow Thunder is ▦ Orange Fire is ✦

Red Lake is ⊕ White Heaven is ▨ Bindu point is ◈▨

Look at the enigmatic tree on the left. Now let's callously chop off a few branches to compare the number and image versions. Here is the topmost pair of branches with a trunk of bindu points acting as divider for all the hexagram fractions—the fractals, in other words.

```
20202020
• • • • • • • •    is
55441100
```

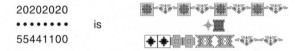

In both versions, you sense a rhythm of 2-ness in the branches. The overbranch has an *alternating* rhythm by 2s, but the under-branch has a *pairing* rhythm by 2s. Notice how both the number and image versions have the same form, but different specific contents. In the image pair of branches, the rhythm of 2-ness is gracefully vis-ible, but its counting function hardly appears at all. For example, you can't subtract ⁗ from ▦ and get 2 . . . that is, unless you know the legend and realize that ▦ is 2 and ⁗ is 0, so that in this subtraction problem, the remainder is ▦ or 2.

Here's another pair of branches appearing halfway down the trunk. At first, its overbranch seems to be a repeat of what we've already seen. But its underbranch shows that this not quite the same setup as we found earlier. This underbranch is different.

```
20202020
• • • • • • •    is
77663322
```

Compare it with that earlier underbranch. This one has different symbols, and it is shorter. But from the analog symbolism alone, you can't tell that according to the linear approach, this lower branch adds up to 36 compared to that earlier branch's 20—much more!

Let's reunite these branches now within the whole structure. Consider our classical tree zooming up laden with RNA message hexagrams. To me this variety of tree looks like it's a hybrid of abstract symbols that come from playing cards or maybe an old alchemical manuscript. It has a formal aspect that is reminiscent of symbolism that has become reified into the heraldry of the past.

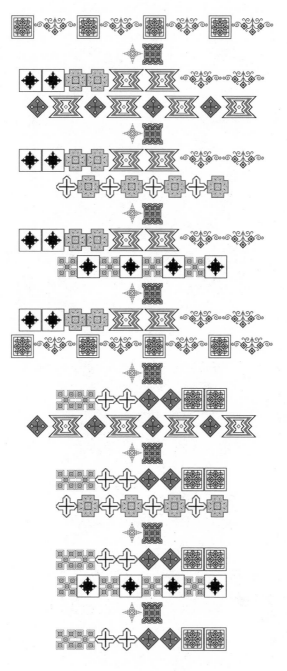

Classical Tree

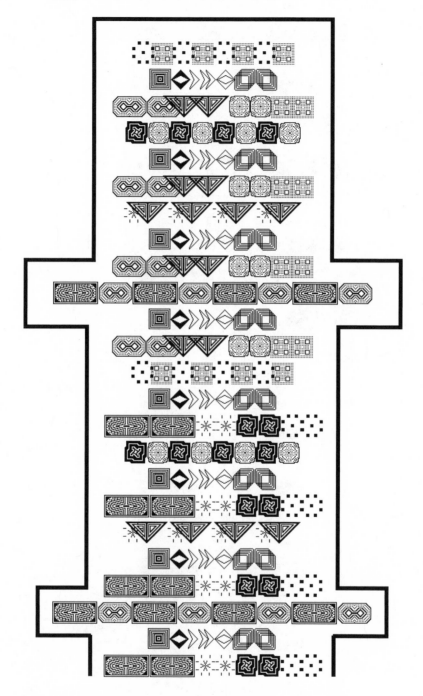

City Tree

This next tree grows in the squared-off stone and asphalt jungle we have made to replace nature's curves. Take the top branch of—

—it's the penthouse apartment garden, also known as 20202020. I like the way these images interleave with each other, rather like . . . well, leaves. In some places, blossoms seem to burst forth, as with—

—also known as 66 or 33 or 22. Sometimes images only peek forth from the overlapping leaves, as in—

—and if I stare at this branch long enough, I begin to see jaguars on it. Never do I see 64646464, which is their numerical version. The animal archetypes still lurk within the human psyche. They re-emerge as projections of totemic wisdom and cunning and strength peering out from the underbrush of life.

Owls also hide along this ledge of city stone. See that one nesting in the cubby? Its other name is 5, and it is surrounded by chunks of 7 and 7, while 1100 is over its head. The mate 5 sits on guard nearby.

And look at those bindu row of apartments! With people who are talking and looking outside.

Overwrought imagination? Well, visualize people walking around a little gallery in New York, stopping before this image to discuss the statement it makes. "Hmm, *City Tree* . . . I like its stark yet textured quality." What if the placard says: "This graphic shows the genetic code. It is also the I Ching." Viewers might nod wisely and yet be none the wiser, for being so literate, they wouldn't take it literally.

City Tree

Black Earth is ▓   Purple Mountain is ◉   Blue Water is ⦂⦂

Green Wood is ▦   Yellow Thunder is ∞   Orange Fire is ▽

Red Lake is ✳   White Heaven is ▣   Bindu point is ▣◈〉〉〉◇▯▯

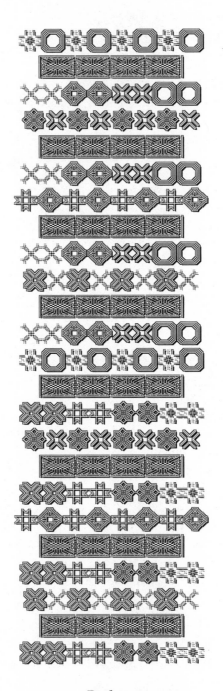

Basketry

Basketry

Black Earth is ▢   Purple Mountain is ▨   Blue Water is ▦

Green Wood is ✾   Yellow Thunder is ◈   Orange Fire is ✕

Red Lake is ▦   White Heaven is ▦   Bindu point is ▨▨▨▨▨▨

This eight-tiered Basketry holds the RNA message hexagrams that create the cornucopia of living matter. Let's examine a few, removing them one at a time from the basket weave of bindu points. This weave was a gift from synchronicity. It came when I hit the wrong key on the computer.

Here is Hexagram 35, *Easy Progress*, removed from the fourth row down of bindu point weave on the right:

It's the hexagram for Tryptophan, the amino acid that you can take for relaxing naturally, something you've experienced, even if you've never heard of it. That's why people doze off after the big holiday dinner at Thanksgiving. It's not the size of the meal that knocks people out, it's the amount of tryptophan in the turkey helping.

These trigrams in *Easy Progress* are Orange Fire over Black Earth. The design might be taken as rays of sunlight beaming over the four-square earth, which is a pleasant notion. It gives a visual metaphor for the sun rising in its natural, easy passage above the earth. But I didn't plan this particular symbolism for Hexagram 35. The computer mix just came out that way. I developed these symbols merely in terms of their image width, so as to get a less branched form. I also tried to alternate darker with lighter, looser with heavier forms for the fruit and flowers inside the happy accident of the basket, as you can see by studying the legend.

It soon becomes evident that the Arabic ciphers of 1, 2, 3 do not emphasize analog qualities. For example, 2 does not really show 2-ness, not nearly so well as that old Roman numeral of II. Of course, in any symbolism we do lose some aspects, but usually it's the analog aspects that go. Why did we switch to Arabic ciphers? Because it proved so extremely useful for adding, subtracting, multiplying, dividing—all those linear tasks of finding quantity. But what about quality? The qualitative aspect of 2-ness is left out of linear tasking.

Nor does this symbol of ❋ , which stands for 2 in our legend, really convey 2-ness in linear terms. Removed from its basketry, it even loses its analog function, for this image ripped out of context cannot show its place and proportion in the whole design. But as a bit of the hologram, ❋ does still convey the quality of 2-ness within itself. Notice how its design is composed of two parts which cross each other, making four polar extremes—and in the center is that holy temenos of the uniting fifth spot—the *quinta essentia* that connects its four poles. This network of connection expresses the qualitative nature of ❋ rather than its quantitative function.

So although we cannot add up ❋ and ❋ and get 4, we add ❋❋❋ and get visible connection. Or combined this way—❋ —we get Lake over Lake, Hexagram 58, *Shared Joy*. It depicts the image of lake feeding into lake and thus enlarging the whole.

Nature itself is stippled with the subliminal number code. When we look at a tree, we don't see numbers airbrushed into its surface, only the counter-levering limbs and ballooning foliage and color. But analinear number exists submerged in the texture of all this, spinning its iterating fractal relationships that push the forming edges of life and make it interesting.

Image as words, image as design, image as impact. Art is full of imagery that resonates at a level so deep that it taps into the vast wheeling pattern beyond linear consciousness. The West has long recognized the Orient's special sensitivity to image and design. We can see it in calligraphy, in paintings, in the old architecture, in clever tools and subtle silken garments. Such attunement bespeaks something more than mere good taste. It shows analog awareness of "what likes to go together." The I Ching itself is 64 clusters of "what likes to go together." We are very fortunate that the ancient Chinese were able to perceive and hand down to us this amazing vision that connects analinear number with philosophical meaning. As the ancient song that is found on page 113 states, we do indeed owe a great debt of gratitude to those who discovered this cosmic paradigm and to those who passed it down—such as King Wen, the imprisoned recorder of the I Ching text.

Finally at the end of this chapter, I offer your imagination a few more trees. Make what you will of them and their play of associations. As for the titles, don't forget that in the analog realm, the pun, being the lowest form of humor, is also the deepest.

Mystery

Miss Starry

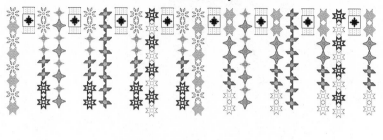

Rocketry

Pagentry

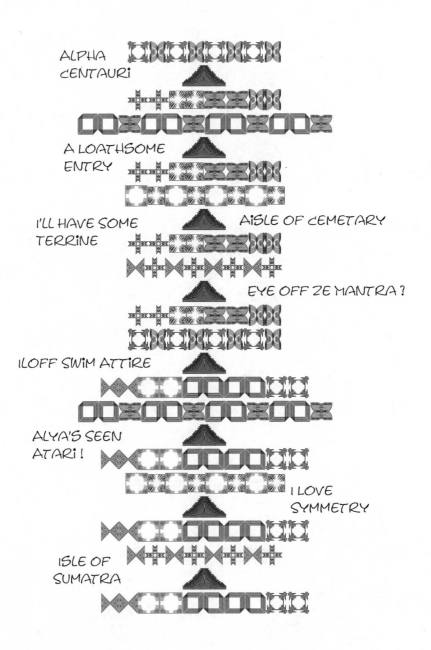

ALPHA CENTAURI

A LOATHSOME ENTRY

I'LL HAVE SOME TERRINE

AISLE OF CEMETARY

EYE OFF ZE MANTRA?

ILOFF SWIM ATTIRE

ALYA'S SEEN ATARI!

I LOVE SYMMETRY

ISLE OF SUMATRA

Save a Tree

Reflectory

PLAYSHOP 10

Plan a program that combines discussion and experiences. Select book paragraphs to introduce the following topics—or your own.

Opening ritual. Announce this session is Playshop, not Workshop.

❧ Ask someone to read aloud the poem "The Pleasures of Merely Circulating" on page 191. If possible, have a drum accompaniment. Do not worry about understanding the poem so much as enjoying it. Ask someone to sing the first verse of the poem to the tune of "Ninety-Nine Bottles of Beer on the Wall." Have everyone join in and sing it in a lively manner several times.

❧ Photocopy the picture of the *City Tree* on page 220, blowing up its scale until the image almost fills the sheet of paper. Now make enough copies for everyone in the group. At the session, pass out the sheets and ask members to sign their names *on the back, not the front of their sheet*. Pass out colors and ask members to apply them in any way they see fit. They may color in the designs, draw scenery, people, more code—whatever. Afterwards, collect the sheets, hang them up with tape and have an art show, with members strolling from picture to picture. Ask the group to vote on the brightest, wildest, calmest, and most appealing. (A single picture may qualify on several grounds.) Ask winners to stand in the center while members applaud them as artists who are honored *inside* the social circle. Dub each into the Order of Art du Gene-Crepe de Chine. Afterwards, talk for a few minutes about what the group members noticed and experienced during this exercise.

❧ Forefinger dance of symmetry (to background music: Look at your left forefinger. Now look at your right forefinger. How are they different? Similar? Get up and stroll around, asking to see the right or left forefinger of each person you meet. Show your opposite forefinger. Next, continue, but without any words, only gestures. After several minutes of silent show, begin to hook your forefinger with the opposite forefinger of each person that you pass, so that people are doing the do-si-do around the room using fingers. Finally, sit down and discuss what this exercise in complementary symmetry shows you. What tendencies appear?

Closing ritual. Five minutes of feedback. Announcements.

The Atomic Map

Recently I had a dream. It was a long, lovely, wordless ballet of yin finding its own true identity through mirroring in sequences of opposite and likeness. There were four movements to the dance onstage, shown in this diagram. It began with the black box on your downstage right. Yin flowed clockwise through the quartering relationships in a mirroring ballet. It was a dance of yin discovering that "I am this . . . and sort of like this . . . not at all like that . . . and not exactly this. But this, yes!" After the reflective exploration, yin finally returns to her original position, now knowing her identity, aware of herself compared to other: "I am only and exactly and completely this."

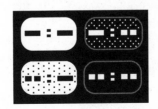

Yin Ballet

This dance created the four bigrams. They parallel the gene's paired pair of T-A and C-G. Both are further polarized polarity.

Old truths come to us in new guises. As we have seen, the four molecules of T, A, C, and G can arrange themselves into only 64 combinations along DNA. This double helix is a stable archetypal form, yet its length is strung in an infinite variety of evolving contents that carry the determined chaos of life. Number weaves a firm analinear grid to maintain the supersystem. Whether as I Ching or genetic code, it is the polarized polarity, a pivotal 4-ness beyond simple opposition that allows the master code to swing both ways, binary and analog.

And now I'm going to show you something remarkable: these four genetic molecules even show up on an old I Ching document called the Ho Tu.

Impossible? Then wait and judge for yourself. Early into my study, I found this surprising feature. Its inspiration carried me on through the remaining years of worksheets and computer tables and German and American text versions to the book that you are reading now. This discovery seemed touched with the numinous. It beckoned me onward like a beacon.

Even though I knew it was risky. The numinous is a stage beyond the luminous—it is somehow touched with divine energy and radiance beyond our limited consciousness to comprehend. We sense its illuminating presence in our lives for a moment and can never call it back at will. But once we have experienced the numinous in an event, we are forever touched by its beauty and revelation. History holds many such stories, but right now I think of the chemist Kekulé, who felt the numinous when he dreamed of a snake holding its tail in its mouth to make a circle. That dream inspired him to realize that the benzene ring is a circle of connective atoms, and that gave the key to organic chemistry. As line becomes circle, linear becomes analog and begins to hold life. Often creativity is literally inspired . . . by an energy that is breathed into us beyond conscious knowing, bestowing an insight beyond the linear data. We breathe in this spirit and get in-spir-ation.

But the danger in the numinous was thought-provoking to me as I read and pondered, for example, Martin Schoenberger's book *The I Ching and the Genetic Code*. Obviously the germ of a big idea was in there, a profound pattern of cosmic connection. He felt it, saw it, printed some portion of it onto paper for us to see. (Marie-Louise Von Franz had first remarked on it, but she never pursued it beyond a brief mention in her essay, "Dialog uber den Menschen.") As I read Schoenberger's book, I felt his love for the topic and his dedication to it. Obviously the whole subject was numinous for him, and I could well empathize, since it felt that way for me too.

I felt, however, that sometimes Schoenberger's ardor muddied his clarity . . . which is often the case with numinous material. In fact it is a special problem for people investigating the boundary between science and mystery, whether it be in the black hole physics of these days, or medieval alchemy six hundred years ago, or the emerging Greek geometry of twenty-six hundred years ago. The closer you get to an emerging paradigm, the more you can become overwhelmed by its grandeur, lured, fascinated, lost in the rush of awe. There is a dangerous enchantment of the deep—the mesmerizing drowning depths—whether it's in the briny ocean, in the well of black space, in the infinity of fractal windows that open up on the computer screen, or in the unplumbed depths of the creative unconscious.

So I began to make a special effort to stay clear-eyed about this material, recognizing that its numen could cast a searing glare or bestow a clarifying perspective. The old myths tell us of mortals who looked on Jehovah or Zeus or other gods and were blinded or burned to cinder or turned to stone or to something less than human. These myths have a psychological root in truth. This numen is awesome. So I wanted to be very careful to stay alert and wakeful, humble and guided . . . not turn away or to stone or inhuman.

I'd already seen enough of the chaos structure in the I Ching and the genetic code to intuit that their fit would probably reveal an ana-linear dynamic, although I couldn't as yet support any hypothesis about it. But already I saw many features in the spinning 2s and 4s and 6s and 8s with their lovely symmetry and fit and resonance, both physically and metaphysically

I was also getting brain overload, wearied by so much invisible work. The pursuit was turning so—so yang, so full of constructs of lofty theory! So intellectual and remote.

So one day in April of 1986, I sat down and decided to search for something yin in this fit between the I Ching and genetic code. What did I mean by this? Well, something just lying out there, silently waiting to be noticed in the background. Yin is opaque mystery, the mute matrix at the background of attention. Silently, patiently it holds, relates, contains. It is the opposite dynamic to yang's linear reaching out, attracting attention, demanding climax and resolution. I wondered how to look at the background without pulling it into foreground. Somehow I felt that I'd neglected something deep and silent in all these coruscations of changing lines and overlaid coding and so forth. I'd missed something that was the passive container of it all. It lay dormant and awaiting my attention. I felt it holding its silent truth ready, if only I could just pay heed. And it whispered somewhere in the wisdom of mute matter.

What was calling me? I could not have said at the time. I did not know. Something dark and wordless was pulling me toward mute evidence in matter. It was a corrective for studying all this ever-loftier bright gridwork of co-chaos charts and theory. I missed the calm, grounded, unspoken backdrop of the great dark feminine. The truth beyond words hidden in mute matter. *Mater,* mother, the *prima materia.* The rich dark mystery of us all exists in this matter.

I said to myself, "Look, if there's so much agreement coming from this analysis of the bifurcation structure, surely there must also be evidence of another sort. In some material artifact or whatnot."

What existed as a material artifact? Not much. Two old maps that had been handed down for thousands of years, called the Ho Tu and Lo Shu, protected in that ritual way the Chinese so long maintained to show reverence for the stupendous depth of their culture. Forgetting even exactly why. Even going "modern" finally and repudiating it during the Red Shift and the Cultural Revolution.

So I got out various books and looked at these two old maps. The Ho Tu and the Lo Shu. Just black and white dots in groupings, sometimes drawn with lines connecting them, sometimes not. You can find them in the translations by Richard Wilhelm and James Legge, but some more modern versions ignore them as not important.

Yet these old maps are the crux of the I Ching. According to the ancient text, the whole mechanism of the Changes developed from these two documents. Yet each map is no more than a cluster of black and white dots in a pattern.

The Ho Tu means the "Yellow River Plan." Legend says that in 3322 B.C., Fu Hsi saw a dragon-horse rising out of the Ho River, and from the spots on its back, he got the idea for arranging these black and white dot clusters into a plan or map you see on the next page.

Now your idea of what a dragon-horse might mean is as good as mine. I've made many speculations on that dragon-horse, ranging from the wholly psychological to the purely physical. Psychology might say the dragon-horse is a creative symbol of libidinal energy rising from the watery unconscious, channeled by the symbol of the Ho River into a directed, applicable, and useful form. This hybrid beast—half mundane, half divine—combines the go-power of a horse with the dragon fire of inspiration. It was energy that rose from the channeled watery depths to bring Fu Hsi his inspiration.

And as I ranged all the way across the possibilities, I came to a completely literal other extreme. One could argue that an ancient mind might describe a spaceship as a dragon-horse because it emits a plumey fire like a dragon and carries things like a horse. You could posit that the I Ching procedure was an alien gift coming from a culture far advanced beyond ours. That notion, however, would only push back the question of discovery to a greater remove. It still doesn't explain how anyone—human or otherwise—actually tapped into the subtle philosophical level of this co-chaos supersystem.

But be that as it may, the original map was preserved until about 1050 B.C. before being lost. According to James Legge, " . . . the thing, whatever it was, is mentioned in the Shu as still preserved at court, among other curiosities in B.C. 1079." Then after it was some-how lost, new copies were made from memory, and despite the vicissitudes of history, they have been handed down to the present.

Ho Tu

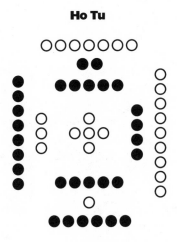

I wish the Chinese would discover an ancient untouched burial site that contains the Ho Tu exactly as it looked at the very beginning. I suspect that our own version may have been stylized over time. It's likely that the number of dots is the same—since number is relatively easy to remember—and also their general locations—since placement is recalled by approximation. But during redrawing, perhaps the dot positions shifted slightly to become more stylized. Anyway, this Ho Tu carries the plan, according to its Chinese name.

The second document was the Lo Shu, meaning the "Lo River Writing." According to classical legend, it came from the Lo River about a thousand years later, around 2200 B.C. The great flood-control Emperor Yu was busy inspecting public works when he saw markings on the back of a tortoise rising out of the Lo River. From the markings, he realized how to interface a new dot grouping with the already-ancient Ho Tu and thus cause the I Ching structure to become dynamic, more than just a plan.

Nowadays, many look at the Lo Shu and say, oh well, that's just a magic square. Sure, count the dots and you'll see that their black and white patterns make a magic square of odd and even numbers. But I suggest that it is more than this. After all, the ancient text says that the Ho River produced the I Ching's plan, while the Lo River produced its writing.

Lo Shu

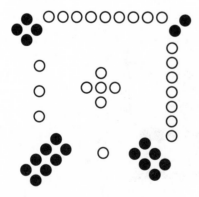

Magic Square

4	9	2
3	5	7
8	1	6

One day I found myself staring at all these dots and thinking, "Now if the Ho Tu is the plan and the Lo Shu is the writing, how does that parallel with the genetic code?" Immediately it sprang to mind: "Oh yes,—DNA and RNA." After all, DNA is the stable plan locked in the double helix of the genes that we inherit from our ancestors and transmit on down to the next generation. It parallels what the ancients said of the Ho Tu—that it gives the plan.

But it is RNA that does the dynamic writing. It literally *transcribes* the plan—even by using modern genetics lingo—into the building blocks of protein. And the ancients said the Lo Shu gives the writing.

Sitting on my living room couch in Zurich, studying these two old maps, I decided tentatively to view the Ho Tu as the static plan of DNA and the Lo Shu as a schematic of RNA's ability to translate the plan into action. Then I started looking for physical correlations.

Between dots and molecules? Well, yes, just playing around, you know. Relaxing, getting away from ordinary logic, just looking into a couple of old maps as a record in matter from the oldest times, far older than King Wen's hexagrams. I asked myself, "So in DNA, exactly what comprises the basic plan? Well, of course—it's the four molecules of the basic pairs: T-A and C-G. In various sequences they lay down the plan of DNA."

So I studied the chemistry diagram that showed the 55 individual atoms joined as base pairs of T-A and C-G.

Base Pairs of DNA

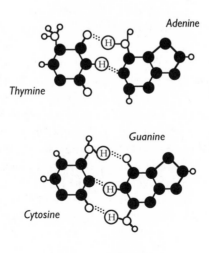

I have labeled the five central hydrogen atoms with a very visible H because that's what bonds each pair of molecules together: two hydrogen atoms bond Thymine with Adenine, and three hydrogen atoms bond Cytosine with Guanine. This bonding is shown by those five dotted paths.

But what do these molecules have to do with the Ho Tu? Would there be an answer waiting silent, unnoticed in mute matter? On a worksheet I drew the Ho Tu from Da Liu's *The Evolution of I Ching Numerology*. Then from page 132 of Watson's *Molecular Biology of the Gene*, I photocopied a chemistry diagram of the H-bonded base pairs of T-A and C-G. You can see that worksheet on the opposite page.

I arranged my worksheet so that the modern molecules sat just above the ancient Ho Tu, juxtaposing these two maps together in such a way that my eyes could easily travel back and forth between them. And the first day, nothing at all occurred to me.

But that night when I went to bed, I dreamed that god picked me up and carried me . . . somewhere. Just where, I cannot say. I only know that the trip seemed to go on all night and it was wonderful beyond words to tell. I woke up in a blissful composure.

That morning in Zurich I carried my worksheet to a Jungian talk already on my week's agenda. It was a week-long series on the symbolism of the Divine Conjunctio, the ultimate union of all things, and sitting there in the audience, I did an odd thing. I put the worksheet before me and gave the lecturer only half my attention; the other half went to my worksheet. With one ear and eye and half my brain, so to speak, I heeded his talk, but the other half studied the worksheet.

Sounds split, doesn't it? Well, now I think that maybe this split attention acted rather like a double induction process in hypnosis: it allowed unconscious material to come up into awareness and be integrated. Here's what happened in that audience with the lecturer buzzing—very interesting, really, but droning on and burdening and hypnotizing with tasks the workhorse of my conscious ego so that the unconscious dragon of creation could surge forward and fly.

Look at the worksheet. Notice that in this fancy version of the Ho Tu, you see some lines drawn between the various dots. You also see the compass directions and the number totals for each group. It is an "explicated" version of the old map, and Da Liu makes it clear that the original version had no such lines or labels, but only the dots to show the pattern of 55 spots "on the back of a dragon-horse."

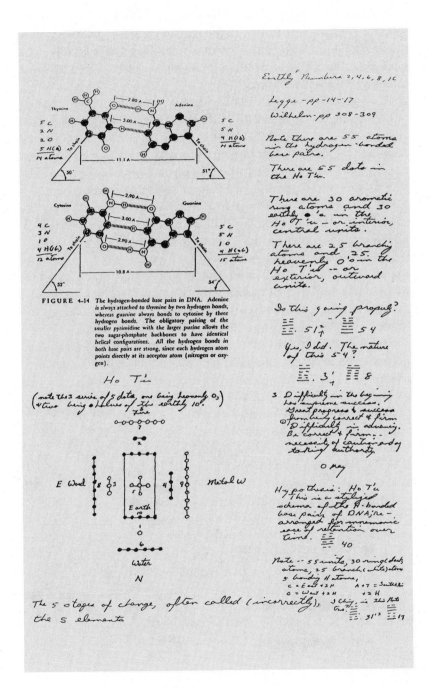

FIGURE 4-14 The hydrogen-bonded base pairs in DNA. Adenine is always attached to thymine by two hydrogen bonds, whereas guanine always bonds to cytosine by three hydrogen bonds. The obligatory pairing of the smaller pyrimidine with the larger purine allows the two sugar-phosphate backbones to have identical helical configurations. All the hydrogen bonds in both base pairs are strong, since each hydrogen atom points directly at its acceptor atom (nitrogen or oxygen).

"Earthly" Numbers 2, 4, 6, 8, 10

Legge - pp - 14 - 17
Wilhelm - pp 308-309

Note there are 55 atoms in the hydrogen-bonded base pairs.

There are 55 dots in the Ho T'u.

There are 30 aromatic ring atoms and 30 earthly •'s in the Ho T'u -- or interior, central units.

There are 25 branching atoms and 25 heavenly O's in the Ho T'u -- or exterior, outward units.

Is this going properly?

51 [?]   54

Yes, I did. The nature of this 54?

3 [?]   8

3 Difficulty in the beginning has supreme success. Great progress & success from being correct & firm. Difficulty in adversity. Be correct & firm -- necessity of caution in day taking authority.

O Kay

Hypothesis: Ho T'u This is a stylized schema of the H-bonded base pairs of DNA, re- arranged for mnemonic ease of retention over time. 40

Note -- 55 units, 30 ring(dark) atoms, 25 branch (white) atoms, 5 bonding H atoms,
C = East + 2H   A + T = South & N...
G = West + 2H   + 2H
J Chi ... in the Note true. 31 [?] 17

Ho T'u

(note this series of 5 dots, one being heavenly O, & two being • halves of the earthly 10.)

Fire

o o o o o o o
7

E Wood 8 3   5 4 9 Metal W

E arth
10

Water

N

The 5 stages of change, often called (incorrectly), the 5 elements

Perhaps it was just as well, though, that I made my worksheet with that fancy lined version of the Ho Tu instead of just the 55 bare dots. It was those added lines, perhaps, recalling the H bonds of the base pairs, that gave me a wild notion: maybe the Ho Tu showed the atoms of the four base molecules.

Absurd? Well, I was just playing, letting the dragon try his wings.

I began to think, "What if I had never seen a chemistry diagram before? What if I viewed this Ho Tu as base molecules, never having seen any molecules diagrammed before today? What if I'd never heard of the usual way to portray hydrogen-bonded base pairs? Get simple. Get stupid. Look at this map with the natural mind."

I got so stupid, in fact, that like a child, I even stooped to the mere counting of dots. First I counted all the dots in the Ho Tu—55. Then I counted all the atoms in the H-bonded base pairs—55!

A shiver went through me. How many things in the world share the number of 55 units? 55 is a peculiar number too, not rounded off by twos or tens. Nor is it typically found in folklore, like 3 or 40 or 7 or 9. Mere coincidence, you say, this 55 that was silently calling my attention to the old map and the new chemistry diagram?

Perhaps you need to count the dots too, just to believe me. This diagram places the two maps side by side so you can do it easily.

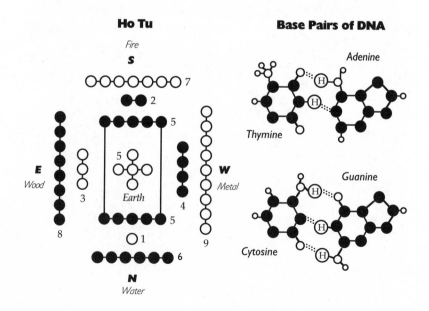

I had to count them several times myself . . . 1, 2, 3, 4 . . . just to believe it. You can see the resulting notation in the upper right side of the worksheet:

> **Note there are 55 atoms
> in the hydrogen-bonded
> base pairs.**
>
> **There are 55 dots in
> the Ho Tu.**

But I jotted down no notation for the little thrill that was going up my spine. What was this curious number correlation between old maps and new molecules—the 55 dots and 55 atoms? After all, I was now looking at the *atomic* level—as opposed to that molecular level of earlier chapters that had strung base molecules into code, or to the subatomic level where electrons cloud and protons gel into wells of probability.

Where did this lead? Anywhere?

Again I studied the Ho Tu's 55 dots. What, I asked, is the next level of differentiation past mere count-'em numbers? Why are these 55 Ho Tu dots further subdivided into groups of black and white?

Well, black stands for yin and white stands for yang. Some kind of polarity might express itself here that would also apply to the atoms. But what? How? I tried various schemes for polarizing the count of hydrogen, oxygen, nitrogen, carbon. No go.

"Get simple. Get stupid. Look at this map with the natural mind. What in these 55 yin and yang dots of the Ho Tu is polarized like the 55 atoms?" The lecturer's words whizzed past my ears.

"Well, yin means holding, containing, passive doing. Yang means reaching, stretching out, active doing." Then as I stared at the atomic structure, suddenly I saw it! Of course. Inside each molecule, there is a central ring of atoms that holds it all together in a yin-like way. This ring of atoms (that same ring made famous by Kekulé with his snake dream) gives the molecule a central hub, while the rest of the atoms branch out from it.

So with a highlighter pen from my purse, under the buzz of the speaker's words I colored in all the ring atoms a fluorescent pink— you can see that it shows in the book as a dark tone. Then I began to count. The result is noted on the right side of the worksheet.

> There are 30 aromatic
> ring atoms and 30
> earthly black dots in the
> Ho Tu—or interior,
> central units.

The 30 black dots equal the 30 ring atoms! Now for the white dots and branching atoms. You also see that result on the worksheet:

> There are 25 branching
> atoms and 25
> heavenly white dots in the
> Ho Tu—or
> exterior, outward
> units.

The 25 white dots of course equal the number of branching atoms—25. So the 55 atoms are polarized into yin and yang, as either holding or outreaching atoms . . . in exactly the same ratio as the 55 Ho Tu dots!

This time more than a little shiver went up my spine. Total shock! Its intensity almost scared me. I sat in that lecture stunned. The earlier paradigm I'd been working on had only compared the I Ching structure with four bulky *molecules* in 64 patterns. But just playing now, I'd found this new correlation not at the level of molecules, but in the *atoms* of the molecules. I thought, "This shared mapping suggests a parallel of a very different order. It takes things to a whole deeper level—to the atomic level. Here is no longer the dynamics of chaos theory as seen in molecules. Somehow this Ho Tu diagrams the very atoms themselves! This mute map shows their atomic patterns!" I was no longer viewing the gene molecules. Instead, I was reaching down into their *ur*-ground and finding the atoms too are parallel.

Perhaps you can imagine my feelings. What were the odds of this being mere coincidence, after all the correlations I'd already found at the molecular level? Maybe like discovering that an intricate and ancient mural on Ayers Rock in aboriginal Australia equates to the Feynman diagrams of particle paths. The question then becomes, "How did this insight from way back there get to way up here?"

Perhaps it could happen for the paradigm, I decided, through number at the archetypal level. The analinear domain hums everywhere below our attention. So Eastern and Western cultures on opposite sides of the globe could tap into it from two very different approaches. It's like saying that not only is a Feynman diagram like the Australian mural, but they both emerge from the same collective root. In the unconscious, we all are one, we are all children from the same source. Jung's metaphor was that the individual flowers of human minds come and go, but the collective rhizome is buried underground and unites our transitory passages. The root of the I Code is archetypal number itself, buried in everything. That's why the ancient Ho Tu and Watson's chemistry diagram can reveal a shared structure in the mute matter of dots and atoms.

But stare as I might till the end of that lecture, I could ride this brainwave no further. Home I went and gazed at the worksheet until I could walk around and see it in my head . . . without any results. I turned its images this way and that—both physically and mentally. I wondered if maybe I'd carried the notion of going stupid too far.

Maybe I wasn't viewing it quite right. So I asked the I Ching, "Is this going properly?" You can see my question written on the right edge of the worksheet, and the answer too. It was Hexagram 51, *Shock!* with a changing line 2, moving into Hexagram 54, which I call *Delayed Union*. The meaning of the *Shock!* hexagram was obvious to me. I'd felt it all over my body, a shock as electrifying as the doubled Thunder of those trigrams forming the hexagram. But now, as a result, I was *in shock*. I didn't know what more to do, how to take it any further, how to utilize what I'd seen.

The changing line 2 described the quality of the situation:

> Shock comes like a bolt of lightning!
> It feels dangerous, as though
> you're losing your treasures.
> Do not run anxiously in pursuit of them.
> Let go and climb above to the heights.
> After seven days you will find them.

The changing line then turns on into the hexagram of *Delayed Union*. So, reassured, I began to let go of my tension and get above it all. Relax. Let the shock wear off. Give myself a break.

That night when I went to bed, I dreamed that god was carrying me again. Somewhere. But this time the trip was vaguer in memory when I woke up. Satisfying, but without that clarity and simple joy of the previous night. Some unresolved quality still hung about the nightsea journey, something left in mystery.

In the morning I returned to the worksheet. Hexagram 51 said it would take seven days to find an answer. But of course, that was not to be taken literally—analogy numbers do not work quantitatively, but rather, qualitatively. The 7 days merely suggested that recovery would come from another angle of contemplation when enough time had passed. It brought up in me associations with the moon and its waxing and waning cycles. The moon, feminine, relational, tapping into darkness beyond logic. The moon, with its 4 quarters made of 7 days each. Moony, loony, Juney.

What was I doing with this random association? Bringing forth drivel? Sure. But the I Ching's imagery isn't meant to be used logically, but associatively. It calls up related networks of thought, so that it can operate subjectively in your mind within the framework of the hexagram's archetypal form. That I Ching answer was showing me how to approach this conundrum from a different quarter.

On the worksheet I wrote down after the *Shock!* of Hexagram 51, "Yes, I did," meaning that I'd felt the shock, and more particularly, I recognized the emotion tone it described in its changing line.

But what about the upcoming hexagram?—54's *Delayed Union*? "The nature of this 54?" I asked, meaning, what characteristics might I expect of this delayed union? Or in other words, please give me more information to help me facilitate this delayed union.

That answer was Hexagram 3: *Initial Difficulty* or *Birth Pangs*. Its dynamic is born of Hexagrams 1 and 2, the all-male and all-female hexagrams. The resulting birth of the next hexagram shows difficulty in the beginning that has supreme success. Its changing line 1 talks about hesitation and hindrance, but to remain persevering and get help from authority. It turns into Hexagram 8, *Holding Together*. Sure, I would hold together with this effort. Give it my best. Okay.

"Birth," I thought. "It's not like following a cake recipe or a wiring schematic. It's organic, it just happens. Okay, if there's this much correlation, then there's more." So in the midst of my mental labor pains, I wrote down a few notes.

Hypothesis: Ho Tu
This is a stylized
schema of the H-bonded
base pairs of DNA; re-
arranged for mnemonic
ease of retention over
time.

I asked the I Ching to comment on that. Its answer was Hexagram 40, *Deliverance*. The dynamic essentially means going back to an old condition, but with a new perspective. Its core analogy is of slaves who are freed to go back to their former homes, so that now they have a new, enhanced appreciation of their old everyday condition of freedom. Yes, this hypothesis certainly gave me a new perspective on the I Code. And it was going back to the old in a new way. This Ho Tu and atomic diagram reinforced what I already appreciated about co-chaos, but from a new angle of enhancement.

Now, as to getting helpers from authority . . . who might they be? Perhaps some I Ching experts or biochemists. So I opened up books again, looked at chemistry diagrams, studies of the Ho Tu over the last several thousand years. At intervals, again and again, my eyes would travel from atoms to dots to atoms to dots.

Nothing. Seven days, Hexagram 51 had said, a quarter-moon's cycle. I needed to see this from a new quarter of contemplation. Quarters! Hey, there was a certain 4-ness to this atomic diagram. It had 4 units grouped into 2 pairs. So where was the 4-ness of the Ho Tu? How was it grouped into 2 pairs? Perhaps time had moved the dots around in the memory of some ancient scribe doing a recopy; generally, designs do tend to stylize over time. But I couldn't tell.

Then I twigged it. Those directions of N, S, E, and W on the Ho Tu—it's the four quarters of the compass. Sometimes those old scholars divided the map into quadrants. So this orienting by compass would provide a certain 4-ness to the Ho Tu, a grouping of the dots into quarters. But as you can see, the old Chinese style reverses the Western compass; it puts South at the top of the map, North at the bottom, East on the left and West on the right. Oddly enough though, it is the very same format that Western astrology still uses.

So what's next? How do these 4 quarters become 2 pairs?

And then I saw. It happened as I was noticing the Ho Tu's center. There sat a group of 5 white Earth dots. Likewise, in the middle of the chemistry diagram sat 5 Hydrogen atoms. Shivering, I began to play with associations. The I Ching calls these 5 white dots *Earth*. But that's odd, because normally Earth is black, symbolizing yin, while Heaven is white and symbolizes yang. So normally these Earth dots should be black, not white! Several thousand years of Chinese scholars have puzzled over why this central "Earth bond of 5" on the Ho Tu uses white dots instead of black.

Another question is, "Why 5 instead of some even number?" Yin is normally shown as an even number, so that it can bifurcate internally, can pair and relate from its analogs inside. You can see that trait evinced everywhere else on the map: all the black dots group into even numbers, but all the white dots group into odd numbers.

Hey, look! In the atomic diagram, it is the central 5 of the hydrogen atoms that bond the pairs of A-T and C-G. So why not parallel them with the 5 center dots in the Ho Tu? The "Earth bond of 5" joins the four directions, and 5 hydrogen atoms bond the four molecules. So here is yet another correlation between the two systems!

Ah, it also explains that old mystery puzzling the scholars! If these 5 dots at the center of the Ho Tu actually symbolize a universal math principle that includes the 5 hydrogen atoms, it would explain why the dots are called feminine yet are 5 in number, why they are masculine white even though they hold like Earth. Simply put, they demonstrate the core paradox of all conditions—for instance, of life in the midst of death and vice versa, of mythology's humble companion d-o-g leading humanity to almighty g-o-d beyond the dark River Styx, of light needing shadow to exist and vice versa, and so on—all the countless polarities and paradoxes. Since these 5 H bonds combine the outreach feature of the white yang atoms with the bonding feature of the black ring yin atoms, they are *both* yin and yang, *both* analog and linear. They exhibit the central Divine Conjunctio of number that brings the two worlds together.

About this I felt quite satisfied, because it explained—at least to me—a paradox of several millennia's duration, the ancient mystery of the "central white Earth bond of 5." Okay, it must be a basic math principle that is also exhibited in the 5 H bonds of DNA's base pairs.

But there is more, I told myself. If there's this much correlation, then there's more. So I kept looking. Let each molecule stand for a compass quarter of the Ho Tu, I said. But which molecule equates to which quarter? I tried every possibility, I think. Pairs crosswise, pairs adjacent in every quadrant of the compass. I evaluated the options by logic, intuition, final fit in the paradigm, and congruence with old Chinese precepts such as the constructive and destructive cycles of elements. I won't tire you with every permutation and ramification, but only offer some highlights and my final pick.

I turned to the Ho Tu and got stupid again . . . counted dots, in other words. The grouping pattern goes like this:

Ho Tu

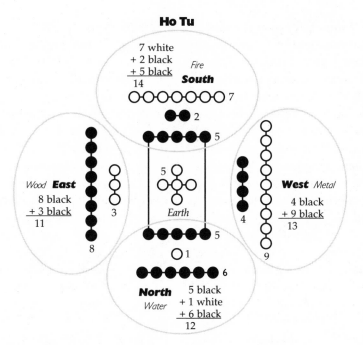

Note the Ho Tu's distribution of black and white dots around the hub of the central 5. Wood in the East has 11 dots, Water *down* in the North has 12, Metal in the West has 13, and Fire *up* in the South has 14. Compare that with the atoms in the base pairs minus H bonds.

DNA Base Pairs

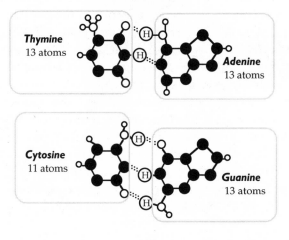

Hmm, once again T and A are the stable pair, each with 13 atoms and also each with 1 hydrogen bond, making a total of 28. And once again C and G are the odd couple, literally, with a total of 27 atoms. These are portioned off unevenly: C has 11 atoms and 1 H bond, but G has 13 atoms and 2 H bonds!

The 5 H bonds become a group of their own, a bonding subset that is like the hub of 5 in the Ho Tu. Excluding this hub/5 H-bonds, the remaining tally is the same: 50. But in the Ho Tu groups, there's a slight slippage of 2 dots that turns it into a counting sequence. One might suppose that after the original map was lost, perhaps a Chinese organizer got busy and "bettered" the layout by a stylization intended to balance the dots more symmetrically in the design.

Ho Tu Dots Minus Hub of 5:		Atoms Minus 5 H-bonds:	
Wood	— 11	Cytosine	— 11
Water	— 12	Adenine	— 13
Metal	— 13	Thymine	— 13
Fire	— 14	Guanine	— 13
Total:	**50**	**Total:**	**50**

But now I recall that to communicate with one another, molecules exchange an atom, just as atoms exchange an electron. It also brings to mind a significant difference between DNA and RNA. The RNA daughter cuts loose and travels with her backbone of sugar-phosphate containing 1 more oxygen atom and also Uracil's 11 atoms instead of Thymine's 14. The upshot is, the RNA system winds up with 2 fewer atoms all told than the DNA it mimics. Although genetics currently doesn't know just why this vital difference exists, I wonder if it might echo the Ho Tu's variance of 2 dots. So I wonder if behind this slight number slippage there is a central truth that might be grasped, given enough time and attention.

I flip the charts this way and that to find a good fit between the specific dots and atoms. T and A don't offer much to work with, since each has 13 atoms plus its H bond, and the pair all told have 28 atoms. The C-G pair, however, offers a useful variation: C has 11 atoms, G has 13; and the pair together with their H bonds contain 27 atoms. This variation may show us a quadrant position for C at least, and perhaps even where its partner G sits on the Ho Tu compass.

One factor in knowing where partners sit on the compass is the ancient Chinese doctrine of the five elements. It is worth considering why there were five Chinese elements instead of the four described in the West as fire, earth, air, and water. Indeed, the Greeks eventually did add a fifth, the ether. Ether was the *quinta*-essential or necessary fifth essence. It brought all the other four into synergy by allowing them to communicate through its essential fifth quality that embraced them all. Ether was a fluid invisible medium, the connector allowing the rest to commingle instead of remaining separate. This ether is the same stuff that, until Einstein, was assumed to fill the universe so that all things could communicate with each other.

The five Chinese elements of Wood, Fire, Earth, Metal, and Water do not stand for elements in the popular Western sense, but rather, they are five phases of change or stages in a process. They interact in a creative cycle and also in a destructive cycle. Below, read each cycle going downward like a long zigzag sentence, so that you will wind up with a constructive sentence and a destructive sentence.

Creative Cycle		Destructive Cycle	
Wood —	*gives birth to . . .*	**Wood** —	*overcomes . . .*
Fire —	*which gives birth to . . .*	**Earth** —	*which overcomes . . .*
Earth —	*which gives birth to . . .*	**Water** —	*which overcomes . . .*
Metal —	*which gives birth to . . .*	**Fire** —	*which overcomes . . .*
Water —	*which gives birth to . . .*	**Metal** —	*which overcomes . . .*
Wood —	*repeats cycle . . .*	**Wood** —	*repeats cycle . . .*

The creative cycle starts productively as Wood makes Fire. Then Fire makes Earth . . . as ash. Notice that this constructive cycle uses Earth as its central condition. Earth appears in the list's center as it goes on to yield Metal ore. Then Metal can turn liquid like Water during smelting. Then Water nourishes Wood, and so the whole productive cycle starts again, iterating itself again and again. You could chant it like a mantra.

Of course, the destructive cycle goes through its own series of phases: Wood depletes the Earth's nourishment; ravaged Earth then muddies Water. Water here in the center suggests the dissolution in this destructive cycle. Water dampens Fire; Fire melts Metal; Metal cuts wood . . . and so on, around and around, iterating endlessly in a cycle of destruction.

But let's not get dogmatically binary and good/bad about this. Destruction is not automatically negative. The cycles of construction and destruction are both necessary for the universe to exist. And most important, these phases called Water, Wood, and so on, are merely Chinese nature analogies depicting processes that are bigger than the words. They indicate mathematical precepts.

How do the five elements relate in the Ho Tu map, and what does that have to do, anyway, with the four molecules? Well, the 5-group in the center of each system acts as its essential fifth condition. This central 5-group is what allows communication among the other four groups. So in correlating this Ho Tu Map with the chemical diagram, let's parse out its 5 central dots appropriately in a way that parallels the 5 H-bonds.

Hmm. Give A-T their 2 H bonds. Now, what would happen with an oppositional pairing of the molecules laid on the compass? For instance, take North-South. Putting South-Thymine with North-Adenine would join the phases of Water and Fire with 2 Earth dots.

Oppositional N-S / A-T Pair

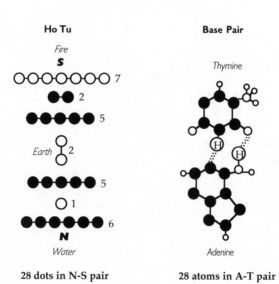

Ho Tu

Fire
S

Base Pair

Thymine

Earth

Water

Adenine

28 dots in N-S pair   28 atoms in A-T pair

But hey, Water and Fire are enemies; they interact in the destructive cycle. Yet in our experience, A and T do not fight—neither in genetics nor in the I Code symbolism, where A and T become stable yang and yin, the "perfect companions." Yet the number totals compute, with 28 dots/atoms in each pair. And there's also a slippage of 2 in the way the dots are distributed—let's hypothesize it's for communication exchange. So this setup sounds as if it fits numerically, but strangely enough, it fits with total dissonance.

The same thing occurs when we make an oppositional pairing of East-Cytosine and West-Guanine, which uses the other 3 Earth dots. Metal destroys Wood. Yet we know that C and G are companions, both genetically and in the I Code, where they become "imperfect pairs." The number totals compute, with 27dots/atoms in each pair. So all told, it seems like this cross-compass equivalence might be a rendering of the destructive cycle of the elements.

Would another pairing of dots/atoms give a more creative and harmonious result? Indeed yes. Instead of setting the molecule pairs into opposition on the Ho Tu, let's make them adjacent. On the next page, we can set up a creative layout where the elements harmonize instead of warring. The number count still works out in the same fashion—still with the slippage of 2 atoms

Oppositional E-W / C-G Pair

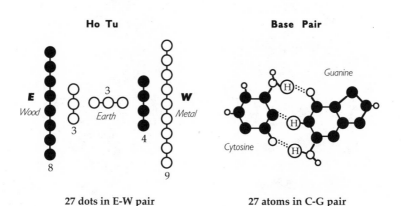

27 dots in E-W pair   27 atoms in C-G pair

This new layout has an elemental harmony to it. Literally. The pair of Wood-C and Fire-G support each other, as do the pair of Metal-T and Water-A. But there are many other subtler correlations, too. We saw that previously, North and South were oppositional, as were East and West; but now we find that the adjacent pair of North and West partake of each other's qualities, as do South and East. It recalls Hamlet's remark that if he's mad, it's only when the wind is from the north-north-west, not when it's southerly. This suggests that compass orientation can offer a useful analogy to the human mind. Things go into balance when grouped SE and NW. Now the purines and pyrimidines are grouped appropriately. Their different atomic rings, of 6s and 9s, also parallel each other. Hmm, 6 and 9. Could this even refer to the 6s and 9s of the ancient yarrow stalk oracle procedure? Or could the slippage factor refer to its 7s and 8s?

Finally, there is a bigger bonus. Now the yang and yin dots are properly distributed between the pairs of molecules. Now there are 15 black holding dots in each pair of adjacent directions, just as there are 15 central ring atoms in each pair of bases.

Adjacent Ho Tu Dots/ Base Pairs

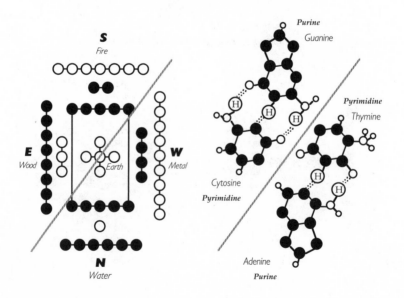

Nevertheless, the slippage or leeway factor still maintains, but it is evenly distributed now, as 1 atom per pair. I don't know why this slippage exists. The deviation may occur simply because the Ho Tu has been shifted and stylized over time. Or it may have to do with communication among the molecules. Or something else.

But I suspect that it's even supposed to be there. To me, it suggests something that is basic to all communication, something inherent in the process. Communication can only occur through exchange . . . whether it's losing or gaining electrons, atoms, time, money, energy. Well, to me, C at Wood/East is the surest match with its 11 count plus 1 from the hub. The other quadrants all have communication buzz going on with the central hub. And oddly enough, there is no C in the four traffic codons—AUG, UAA, UAG, and UGA. Perhaps there is a tie-in here between communication needs and no C.

I also see that the ancient yarrow stick procedure uses this very feature of loss to start the dynamic of communication going. How so? Well, in the base pairs we have 5 H atoms and 50 other atoms. If we just focus on the 50 other atoms, and then allow the loss of one more, we can start a polarized dynamic of communication going. By the same token, in doing the ancient yarrow stalk mutations, we begin with 50 stalks. Then a stalk is immediately set aside "to sweeten the pot," and never again is it touched during the many calculations with the other 49. I see a parallel here of setting aside one stalk or one atom to start communication going amongst the other 49.

The Ho Tu appears to be a stylized atomic map. It reveals a whole new level to the paradigm, different from the molecular level we've explored in earlier chapters. To my mind, this atomic correlation is an independent proof. It's like tallying the city budget, or cross-checking a math problem to see that not only does $7 + 5 = 12$, but also . . . yes, $12 - 5 = 7$. It verifies that the I Ching and genetic code really do spring from the same archetypal co-chaos supersystem.

How did the Chinese come to silhouette the 55 atoms of the base pairs in the Ho Tu? I don't know. Did it come from dreamtime like Kekulé's snake? From some bone-deep archetypal insight? From tapping into an autistic vision of the silent mathematical hologram that embraces us all? From the culture of a lost civilization . . . like Atlantis or Mu or somesuch? From a pit stop on this planet made by extraterrestrials on their way to elsewhere, or shipwrecked and making their missionary best of it here? Or is it sheer chance? Can chance get this sheer and slick and hard to dent? Well, what do you think?

WORKSHOP 11

Plan a program that combines discussion and experiences. Select book paragraphs to introduce the following topics—or your own.

Opening ritual.

❣ Ahead of time, ask if anyone has a bundle of yarrow stalks for doing the I Ching in ancient fashion . . . or buy a pack of bamboo skewers to use. Study an I Ching text—e.g., Wilhelm, Trosper & Leu, or Whincup. Then show features of the ancient procedure to the group. (Don't go through the whole thing unless you can take up to 45 minutes.) Pass around the yarrow stalks if you have them, so everyone can handle them. Discuss. What do people know about yarrow? What is yarrow used for herbally? Where does it grow? Could you grow your own yarrow and use it?

❣ Count off as Wood, Fire, Earth, Metal, Water. Ask the 5 groups to sit in the four compass quadrants of the Ho Tu plus the central fifth position. Now read aloud the destructive cycle of elements several times as a "ball of energy" is thrown from group to group, each element raising its hands to receive the ball when called. What's the ball's pattern? Do it too for the creative cycle. Discuss.

❣ How do you explain the congruence between the Ho Tu dots and the atoms of the DNA base pairs? Does this convince you of a larger underlying pattern? Why or why not? How do you explain it on top of the earlier correlation at the molecular level? How could such a synchronicity have come about? Poll the group and see if you can reach a consensus. If not, what are the main camps?

❣ Think back to a time when you had an insight that was other-worldly or inexplicable from a logical point of view. Describe it to the larger group, or in small groups. Where do you think such insights come from? How do you explain creative moments? Is there any way to cultivate them? Is it worth bothering about?

❣ What is the real difference between noticing synchronicity and succumbing to superstition? Can you train yourself to increase the former and avoid the latter? Or do you think they amount to the same thing? Listen to other experiences and recount your own. What does this range of experiences tell you about reality?

Closing ritual. Five minutes of feedback. Announcements.

Section 4

Lofty Spirit, Humble Mind

Earth

Heaven

11   **Harmonious Unity**

The Judgment
> The petty goes.
> The great comes.
> Good fortune. Success.

The Image
> Earth descends
>   to meet the surge of heaven.
> Their natures swoon
>   in consummating union.
> Mind with matter merge,
>   meld male and female matrix.
> Strength surrenders,
>   soft suppleness surrounds.

Consulting Procedure

Tapping into the I Ching offers you a two-way conversation, not just one-way prayer. Here's a simple, accurate, modern way to consult the oracle. I like this method because it has superior harmony and ease. It is faster and more aesthetically pleasing than the coins, yet it still retains the ratio of change found in the ancient yarrow stalk method, which the coin method does not. I am grateful to Richard Wilhelm and Hellmut Wilhelm for foreshadowing it by their various discussions of the yarrow stalk ratios, and to Larry Schoenholtz for describing it as the "16 system" in *New Directions in the I Ching*.

Essentially, one uses a little consulting kit. Here's how I made my own. First I color-coded the four I Ching lines according to black earth, clear air, red fire and blue water. Each color symbolizes a specific kind of line. *Stable Yin* is black, *Stable Yang* is clear, *Changing Yang* is red, and *Changing Yin* is blue. Then I purchased some flat glass "stones" in these colors. (I call them stones, since they look like colored quartz). They are used in the ratio illustrated below. Notice, to show any hexagram changing line, I mark a dot after that line. Some people use X's and 0's, but it just gets too messy in my handwriting. Still, that must have worked fine back in the days of writing in the dirt.

Stable Lines		Changing Lines	
yin	yang	changing yang	changing yin
7	**5**	**3**	**1**
Black Earth	**Clear Heaven**	**Red Fire**	**Blue Water**

Use **7 black** to **5 clear** to **3 red** to **1 blue** stone.

You can make your own kit by using some small stones or beans or buttons or marbles or beads—whatever small objects you find handy—but they *must* be of the same size and in four different colors. The first day I ever tried this technique, I used gum balls—it was the only thing I could find in enough quantity and color range at the toy store on that Sunday. Later, I went on to employ beads, marbles, and then little pottery pellets that I formed out of four shades of clay, flattened into "stones" so they wouldn't roll around and fired for durability. Finally, though, I've decided on glass stones, for several reasons. Glass is a natural substance—silica. With glass, I can get uniform weight and texture, as well as size. Its black mimics the opacity of Yang Earth, while for Yang Heaven, the transparency of clear glass offers a better symbolic representation of the sky's clear air, in my opinion, than does opaque white. But in a pinch, any objects of sufficient quantity in a range of four colors will do . . . even gum balls.

I placed my colored objects in a container using the proportions of 7 black stones to 5 clear stones to 3 red stones to 1 blue stone. This adds up to 16 stones in all. The ratio can of course be multiplied to put more stones in a container—16 x 4 = 64 for instance—and the stones thus gain more "mixability." But it is not necessary for accuracy. Only the proportions are necessary, not a specific number in the total count, nor are these colors of black, clear, red and blue.

As to the container: it may perhaps be a pouch or metal can, a wide-mouthed vase or wooden box. Personally I use a zippered pouch of just 16 stones, because it is quiet, it packs flat for traveling, and it can easily be tipped to one corner to corral all the stones. This pouch may also be purchased commercially, if you don't want to go to the bother of making your own.

The pouch of stones became my ongoing I Ching set. Whenever I want to consult the oracle, first I ponder my question. Then I write it down, just to be very clear about what I am really asking, and also to keep a record. I don't ask a question that demands a binary yes/no answer. Instead I pose it in a format that allows a qualitative response, something that will show me the *quality* or flavor of the situation at issue. Sometimes I'll write down several sentences to get my concept clear. (For instance, here's a question that I asked yesterday: "I wonder how my upcoming trip to Los Angeles will be and how I will experience it. What's in store for me on this trip?")

After I write my question, without looking inside the pouch at the colors, I draw out one stone. This gives the bottom line of my hexagram. I write down this line keyed to its color. For instance, if it is a black stone, I would write down the first line as stable yin. Then I *replace* that first stone back into the container in order to keep the proportions true. Next, I draw out another stone to get the second line, again keyed to its color. For instance, if it's red, I would write down the second line as changing yang. I do this six times in all, and thus I wind up with six lines, quickly building my hexagram up from the bottom line to the top. This gives the answer to my question.

How to interpret it? Well, you need a book of the 64 hexagrams with their interpretations, so that you can spot your hexagram on the book's coding grid page, and then look up that hexagram in the text to find out what it means. Study its analogy. Examine how it fits with the pattern of your own life. Make a written note of your hexagram and your main thoughts about it, so that in two weeks you won't have to depend just on memory. Memory can be a real fooler, you'll find, when ego gets in the way and protects itself by blurring the issue.

Here's the answer regarding that question about the trip. First I drew from the bag a red stone; second a white; third another red; fourth a white; fifth a red; and finally *another* red. Next, I must find out just which hexagram this is. The coding grid page in any I Ching book will show it is Hexagram 1 ☰: *Active Heaven* . . . and if you look ahead on page 263, you will find an explanation of this hexagram.

Remember, the dots after this hexagram show it has four changing lines. They will turn it into Hexagram 40 ䷧ *Deliverance.*

Uh-oh, I believe that Hexagram 40 is not very well explained in many I Ching versions. I call this dynamic something like *Back to Square 1½* or *As You Were . . . Almost.* It means that a situation is restored to its former condition, but with a certain new knowledge and freedom, so that a subtle gain has been made. There's an element of being delivered from trouble back into an old condition, yet now also with a new scope, so that you feel a certain *deja vu* effect even as you can deal with the old in a better way now. In ancient China, the analogy used was of prisoners being freed back into their old condition before imprisonment. Their "Deliverance" releases them back into their former state, but now it is experienced with a new insight and appreciation. From this example of the "freed prisoner," you can see how complex and nuanced is the energy of a hexagram. It shows a subtle mindset. Its answer is a far cry indeed from a mere flat, dogmatic, binary yes or no.

The I Ching is also necessarily experiential. I can tell you things about it that may enhance your perception of the oracle and its answers, but only you can take the leap into experiencing it for yourself. Only through time—if you bother—can you actually come to realize how it can showcase the evanescent quality of an event, reveal some unexpected insight, or spotlight an all-too-familiar facet in the whirling holographic jewel of your psyche. Each hexagram will describe a unique dynamic, a special flavor in the ice cream shop of reality.

If you're already familiar with the I Ching, you'll notice that some of the hexagram titles that I use are not those made familiar in the best-known English translations. Over time, I have recorded and studied over 20,000 hexagrams of myself or clients relating to real-time events. I've also studied the ancient text in China. When one consults with several scholars, one finds that translation of the characters can be rendered with wide latitude, so it is understandable why I Ching versions vary so widely.

The real issue is to pinpoint the archetypal dynamic behind the words. A visual image can come closer than a verbal image, and a verbal analogy can come closer than a chain of logic. The task is to find the image within the words and then the dynamic within that image which moves it into analogy. It is the silent instantaneous flash of insight that comprehends an archetype.

An archetype will evolve in its manifestations. But the essential mathematical form does not alter, even though its specific contents do. Number is the primal archetype, and it does not change—a 4 can hold spears or computers or cloud castles. The form will hold the varying contents and show them, much as the glass pitcher can display various ingredients and yet shape them all into the same form. From generation to generation and culture to culture, each humanized archetype (say, Mother or Father or Magical Child or Devil) will vary and evolve the specific contents that it shows within various societies.

Since I was not always satisfied with comprehending the eternal dynamic behind the I Ching images filtered through translators, I studied the ancient Chinese text. I soon saw that it too was filtered information, this time through an archaic time and limited written vocabulary and social mindset (especially regarding government and royalty, family roles, and the shadow side). It was partly for this reason that I decided to keep elaborate daily records and slowly develop some modern ways to convey the archetypal core inside each hexagram. Sometimes these titles are the same or nearly the same as you will see in standard translations. Other times they have been adapted to show the core energy of the hexagram. Once I spoke to a Chinese scholar in Germany about this experiential record I was keeping, and he said, "But that is just what the ancients must have done. Kept careful records, seen, and described it for themselves!"

As to ritual and so on, personally I do not wrap the I Ching text in silk or store it on a shelf higher than my head or do the various other preliminary rituals. Why not? Well, for one thing, with over forty versions on my shelves by now, that's a lot of wrapping.

And the cosmos works on the co-chaos principle. It knows my intention already. No need to beguile or prime it with perfunctory ritual. I cannot fool it. In my experience, intention is the real key. I need only to keep my head calm and unafraid and my heart open.

True, ritual can be quite helpful if it is used to clear the heart and head of distractions. But remember that ritual is hollow form; it must be filled with your own sincere contents. No ritual is holy unless it clarifies and focuses your spiritual intent. Insincere ritual is sham. It becomes reified and empty of sincere content, just as the Chinese empire slowly hollowed into a courtly ritual that was out of touch with the people. The Chinese are indeed admirable for passing down to us the I Ching after so many centuries, but it is the ever-changing questions put to the I Ching that keep its form alive and relevant, not any amount of ritual. I recall this and seek the source, not an ornamental reverence around it. Mostly I just try to bring each question from myself alive and open to new perspectives.

I also seek to understand the answer with my whole being, not just my head. I try to hold myself open and ready to meditate on the answer, whether it seems agreeable to my ego or not. If an answer is difficult for me to understand or accept, I ponder to learn from it slowly, and even to remember that if my ego pooh-poohs it as incomprehensible, irrelevant or unlikely, I should wait and see. The ego's linear blinders are amazingly powerful. But by now, I know that the I Ching is looking around a corner of time that I cannot see. I work to befriend this hexagram's dynamic, so that into its archetypal form, I can pour the best contents I can muster. I ask to resolve my inner ambivalence and come toward unity. I am a tiny fragment of the hologram hinting at the huge divine oneness of the universal Tao. I seek to become transparent and let the Tao shine through me.

At night I get my psyche weather report for the next day. Usually it is a hexagram which then changes into a second hexagram. But occasionally it's a single stable hexagram. Whatever, I study it as a thumbnail guide for the tenor of the next day. I know that it is far more dependable than any TV weather forecast.

The next day I use the hexagram as a reference point. Just how the dynamic pattern will manifest itself in my day on scales large and small usually becomes evident only when I stop to consider it, for normally the rush of events sweeps me up and carries me away from the level of symbolic meaning. The detail of life plays out its engrossing scenarios so vividly that they catch me up in the whirl.

The psyche forecast of the hexagram thus becomes a portable meditation point throughout my day. Sometimes I just stop to recall it, to let it help me find a sound footing in the underlying structure, so that I can stand firm in something bigger than myself during a taxing event. It can be weirdly comforting to see that well, yes, today really is *Standstill*. All day. Strangely, it helps me relax and not feel guilty, yet also do my best. I see past the short term of today.

If you don't know the I Ching and want to try it, I suggest that you begin by asking your question on a time-limited situation—a day, for instance—so that you can see more clearly how your answer pans out. Don't overwhelm yourself with a lot of quick questions and answers, for working in the analinear scope of the I Ching can exhaust you with its unaccustomed mental gymnastics at first. The vague boundaries inherent in its analog quality can make you feel groggy and ungrounded. And applying it with everyday logic to tangible matter and bottom-line results can be bewildering. If the ego feels too threatened, it becomes defensive/aggressive and locks into self-justification. It all becomes too much to digest and drives you to retreat back into the conscious ego. Recall that psychiatrist who stayed away from the I Ching for ten years?

The I Ching process taps into the holistic right brain and sends you over the boundary of logic into a realm where normal linear rules don't apply or orient you. Going into meditation on a hexagram answer is fine, but only if you can come back and utilize that right brain message that you found. Practice will slowly develop your psyche's muscles. This exercise is pumping iron for the subtle body. Practice it judiciously and regularly and it will make your psyche strong and supple.

1 Heaven

Heaven

Active Heaven

The Judgment
Vast creative energy.
Penetrating power.
Benefit comes from being
Resolute and true.

The Image
Six dragons of creative force,
Lofty soaring energy source,
Filling heaven with yang power,
Noble and sovereign in this hour,
Freely swim and walk and fly,
Take dominion of the sky!

The Changing Lines
Read this hexagram from Line 1 upward to Line 6 for an overview.
Apply the text of any changing lines. ✌ denotes the most important line.

If all lines change to yin: Each dragon loses its head.
    Don't dread. Good fortune instead.

Line 6   The dragon soars past normal limit.
               Cause for rue rises in it.

✌   *Line 5*   In the sky, bright dragon, fly!
               By sage advice it mounts on high.

Line 4   The dragon poises to take wing,
               From the deep it starts to spring.

Line 3   Daring by day, wakeful at night.
               The way is rigorous but right.

Line 2   Fledgling dragon now appears.
               Sage advice helps if it hears.

Line 1   The dragon's dormant in the deep—
               For now let its energy sleep.

Interpretation of the Changing Lines

If all six lines transform into yin by losing their yang-like head of assertion, this is especially fortunate. The Chinese character for *head* can be translated as a single leader who no longer unifies the whole group of dragons, or as each dragon divesting its own head. Beneficial.

Read the changing lines upward from the bottom line.

Line 6 When it soars beyond its limits, yang energy must fall back into the depths of inertia. This decline, although natural and inevitable, still becomes a cause for regret. Yang transforms into yin, retreating into that quiet state foreshadowed by its birthplace in the first line. An excess of activity must lapse into inactivity as the dragon seed of creative energy once more lies dormant.

Line 5 Active creation soars through the lofty heavens, visible to the world. The dragon is mature, soaring at its height of effective power. Goals are reached; results appear. But be careful: this vital force needs wise judgment to guide its power properly.

Line 4 Having gained its full-fledged strength, now the dragon is ready to play freely across the old domain that once held it passive and embryonic. This yang force poises at the edge of action, ready to take flight in mature, confident mastery of the watery element that bore it (suggestive of the emotions, the unconscious deeps, the womb).

Line 3 Ceaseless activity strengthens the creative powers. Day and night, this yang energy races to and fro, checking and rechecking because the challenges at hand seem so formidable. Insomnia is a possibility here. If so, put it to creative use. This time is difficult but developmental; it initiates the dragon's energy into mature responsibility. Although running this gauntlet of trials is tiring and difficult, the way is right. (The only line that doesn't name the dragon.)

Line 2 The dragon force makes its appearance, creeping into the field of attention. Still shaky at first, this raw power needs wise direction for it to mature properly. Its energy must be directed positively or it could become a loose gun of destruction.

Line 1 Creative energy is still hidden unhatched in the oceanic womb of time. Allow this dragon force to mature at its own rate, not forcing it prematurely into action. Holding back takes deep trust in the creative process, knowing that assertive yang potential lies aborning in what appears to be yin inertia.

Associations with **Active Heaven**

Originating Drive	*Foreground of Attention*	*Seminal Work*
Engendering Energy	*Attention-Riveting Motion*	*Dragon Seed*
Active Force	*Lofty Aims*	*Mighty Zeus*

Your Own Association Regarding this Archetype

The Analogy

Active energy is the dynamic of this first hexagram. Heaven ☰ is piled upon Heaven ☰ so that its brilliance is redoubled. The two all-yang trigrams are full of vitalizing motion. The potency of this dragon rises line by line throughout the hexagram. In the lower trigram, the majestic force hatches and rises from the oceanic deeps to walk upon the land and mature in strength. In Line 1, the dragon lies hidden under-water where its strength is not yet displayed. In the second line, it moves up onto land where association with a role model of wise power brings gain. In line 3, by ceaseless activity it grows in strength, much as a fledgling develops by day and night through constant careful exercise.

In the second trigram, the dragon goes back into the water of Line 4, but now it can disport there with mature strength and prepare to take wing. Then in line 5 it soars into the sky at the zenith of its power. Finally in the last line, the dragon exceeds its limit and must drop back into the yin deep of the unconscious, the unknown. When all six lines change into yin, this is especially fortunate. Yang challenges the earth-bound limits of life and death by energizing the future. It penetrates and alters timeless cycles with its purposeful line of intent. Be firm and correct in using this awesome power. Direct its energizing force with the four guides of love, morality, justice, and wisdom.

Analysis

The dragon is a Chinese symbol portraying the active, tireless, creative strength of yang energy. What is a dragon? It is a mythical, magical beast whose dominion is universal, for it can live underwater, on land, and in the air. It inhabits a realm beyond the ordinary limits. It is more symbol than substance, more image than matter. The United States took the eagle, Britain the lion, Germany the bear; but ancient China, on the other hand, honored invisible energy more than material force and thus it chose the lofty dragon as its supreme symbol. The dragon portrays power that lies beyond the ordinary animal range.

What is the psyche's view of this image? I've asked many Chinese what the dragon symbol means to them, and they invariably say assertive power, strength, potency, masculinity, yang force, in both its positive and negative connotations. Western psychologists have concurred by suggesting that the dragon is a subliminal symbol for male sexuality, much like the snake, but because of its wings, more divine and endowed with mystical power. Western myth has portrayed a dragon as something to be conquered. For example, St. George defeats the mighty dragon and rescues the princess in a battle that suggests oedipal conflict waged to defeat the father's power and win the female.

But to the ancient Chinese mind, the dragon was awesome but not necessarily inimical, powerful but not always malign. With care and wisdom, one can befriend this force rather than kill it. One can tame the dragon and ride its natural cosmic power.

The key is not to divide and factionalize the total archetype into yang fighting against yang in a contest of individual strengths struggling to win some yin goal that often indeed becomes secondary to the conflict itself. When yang battles against yang, the result is destruction. When yang unites with receptive yin instead, it becomes creative beyond imagining. "The dragon phenomenon can create miracles!" one old Chinese man told me. It is this chaos pattern of active, creative energy that Hexagram 1 describes.

Even the most potent energy, though, must come to an end. This is the way of the Tao, for each chaos pattern has its cycle. Commentary ascribed to Confucius says, "The dragon admits only advance and no retreat; only existence and no death; only gain and no loss." Such one-sided behavior is self-limiting finally, doomed to become weakened in the inevitable cycle of change. Chinese commentary put it this way: "The bowstring should not be kept perpetually taut, the bow held continually drawn." Because of its constant advance, the dragon finally becomes vitiated, even erratic when its power is carried beyond its limits. It regretfully drops back into the deep again to regain strength.

How shall we interpret this metaphor of the deep? It is the deep cover that hides the inactive dragon's potential. This may be the blur of daily events that seem to lack a directed creative energy. It may be the veil of time obscuring the significance of its deeper holistic pattern. It may be the oceanic unconscious where creative energy lies waiting, building energy to ascend again in a bright heart-stopping flight before sinking back into the fathomless depths.

Summary: Dragon power is the great originating force that is expressed as motion without obstruction in harmonious and enduring energy. This vast and divine force needs to be used firmly and correctly.

A Modern Analogy for **Active Heaven**

Just before starting to work on this hexagram text, I had a dream:

I see an overview of a beautiful place. Then I'm walking along in it, in the countryside. It is a majestic area with mountains ahead and a lake off to the left. I follow the road. The paving stops; the countryside rolls ahead in dips and hills, rising northward into a mountain ridge. I walk a few more steps and stop by a wooded area on my left. I turn back to gaze along the road that I've been traveling. The forest is on my right side now. Something in there startles me. I look into the trees and see a strange . . . what? Animal? Bird?

It must be a large bird, for it's up in the air somehow. Maybe perched on a tree branch? I look carefully, and then I go closer. Is it an eagle?

But it looks greenish along with the brown. Now it starts flying toward me, hissing . . . glistening, browny-green. It comes closer and I see that this is a dragon! A real dragon! It comes so close that I can see its high arched nostrils, the fangs, the forked, flickering tongue in squared-off jaws. Those striped white and red markings on the head ridges are beautiful! It flies right over me. Glorious! Thrilling! What a blessing! But still, I'm frightened. This dragon is so powerful and flexible . . . it can go anywhere.

I hurry back along the road to tell others, glancing over my shoulder to make sure that I really did see a dragon. And it comes at me again! Hissing! And I verify: "Yes, yes, it really is a dragon!" I see it in profile now, the sinuous body rippling in vertical waves through midair, the colorful boxy head with its bright ridges, the wide-open jaws. Amazing! Wonderful!—but also frightening. Then suddenly there is somehow a chain link fence between us, so I'm safe.

Walking back, I come to a train station where a group of people, European to my eye, are emerging from the train. They stand and spit on each other's faces and heavy coat lapels. Shocked, I say, "Ugh, disgusting." But they seem to be satisfied and rush off to go skiing.

People just behind this group laugh uneasily on hearing my words of shock. So I ask in German, "Why did they do this spitting?" No one answers me. Finally a woman says in good English, "They do it to express their humility. In honor of a famous event that happened here, a vision here."

Oh. Then this must be the spot of the dragon vision? Where is it? Am I dead? Did I go beyond that paved road into death? Or was I just moving into uncharted territory? I hear the woman discussing the vision in another language with those around her. I walk along the road, pondering all this.

This dream shocked and thrilled me. Its energy felt so strong. It shows the archetypal force that the ancient Chinese called "dragon" in Hexagram 1. At first the dragon appears from a camera-eye view. It is a mountain ridge lying buried in nature, reminiscent of Chinese *feng shui,* which calls a mountain ridge a sleeping dragon. Next this force drops down to the individual human level of experience. Any creative person must venture off the paved road into the wilds beyond. At first this dragon waits hidden off to my left, perhaps in that lake whose watery depths suggest a personal version of the great oceanic unconscious. Why on my *left*, though? It suggests the left side of my body, which is controlled by the holistic right brain. But then this force moves into my awareness as I turn and reorient on the path. This brings the creative energy into my right-hand domain of conscious action.

Now I see the dragon in the trees (the bifurcation trees?) It flies closer—such thrilling majesty inspires me to go tell others. But it also evokes fear and doubt in me. I double-check. Yes, it's true. It *is* a dragon! I see the bold profile. Fortunately, though, a chain link fence protects me—suggesting that chains of cause-and-effect logic can be knitted into an analinear netting that restrains the numinous power enough to keep me safe. So I start back to tell the others.

Hmm. That's true. At first unbelieving, I checked the validity of this paradigm through years of daily records. And now I'm writing to tell you what I've seen.

The train? It suggests collective momentum along a preset course. These people emerging from the train station . . . they're Europeans. They are leaving behind their city schedules and the rigid train track for personal freedom on skis in untrammeled nature.

Why Europeans? My own association—and dreams always refract one's personal focus—is that this dream foreshadowed the Munich company, Diederichs Verlag, publishing the German forerunner to this book. For this dream came along *before* they published it!

These dream people spitting—ugh!—said my fastidious ego. It didn't like this part. But the secret language of dreams often uses a dumb show of body function to get its message across. In pantomime, the people literally ex-press their saliva. Spit is the inner solvent, the flowing spirit that allows us to swallow the hard lumps of life. It is the transforming water of life in our humble, shadowy bodies. Through this transformative solvent, the lumps become digestible and nourishing.

These people feel satisfied by reenacting this paradox of flowing spirit within humble human flesh. They exhibit it on their faces and lapels, yet! —where we wear the expressions, cosmetics, and badges of honor that we show prominently to society! Then they rush off to go skiing. Well, the Germans and Swiss and Austrians read this co-chaos paradigm first. They skied among its trees along the white filades of pages when English readers hadn't heard of it, when my agent said that American publishers were dubious whether this book would sell.

Back at the station, I try to understand what's happening. Then the foreign woman speaks English, the presentday global language. She says a vision has occurred here honoring the dragon force. Then she starts interpreting it into her own language for her group. I walk on, pondering all this, even enough to remember and write it down now. Synchronicity has allowed me the opportunity to offer you this dragon imagery for Hexagram 1, *Active Heaven*, applied to a realtime event.

41  Mountain

Lake

Starting Small

The Judgment
Start small.
Great fortune, no error.
Right behavior,
Gain come from it.
How to begin?
Even two small pods
Hold enough to start.

The Image
Vapor rises from the lake
  to wreathe the peak in cloud.
Mist shrouds thick upon the rock
  until a storm grows loud,
And torrents pound the stone to soil
  in bouldered streams quick-plowed.

Mountain drops its rugged height
  in flake by stony flake.
Decreasing unproductive might
  seeds fortune in its wake,
As mountain lowers into sight
  and fathoms bottomless lake.

The Changing Lines
♥ denotes the most important line; ✶ the next most important lines.

✶ ▰▰▰ Line 6 Increasing others increases one too.
  It steadily brings good fortune through
  Enlarging beyond the family crew.

♥ ▰ ▰ Line 5 Benefits flow without even "Please?"
  What luck!—like ten stringed cowries
  Or tortoise shells to the old Chinese.
  Great good luck with effortless ease!

▰ ▰ Line 4 A problem is eased by another's aid.
  Action brings joy and delight unafraid.

✶▰ ▰ Line 3 Remember the rule of decrease for gain:
  If three walk together, one leaves in the strain,
  But walking alone brings a friend in the lane.

▰▰▰ Line 2 Passive giving is a gift. Here action is untoward.
  Passive giving gives a lift yet keeps the giver's hoard.

▰▰▰ Line 1 Rushing off to help a friend? Not wrong but do recall
  That too much offered can harm all.

Interpretation of the Changing Lines
Read the changing lines upward from the bottom line.

Line 6 Here without any personal loss, one can bring gain to others. This is right to do! Persevere toward a goal; every move will further the progress. Helpers will come from beyond the family network. (Originally this meant that the help came through the assistance of an unmarried slave or helper.)

Line 5 Here one is enriched by lucky predictions and value worth ten strings of cowry shells. A cowry is a mollusk. Instead of opening in a bivalve fan, it has a serrated mouth indented into a little shiny mottled shell. In China, these were strung as a form of money. You can see these strings in ancient Chinese art. The exact lucky object is ill-deciphered and translated variously; some read it as tortoise instead of cowry shells. How many tortoise shells make a set? Some say two, some say twenty-one. In any case, such favor cannot be countered.

Line 4 This is a weak, even sickly and inert position for yin in this initiating spot of the second trigram. It needs some yang energy. But this yin line gains help from another's strength (symbolized by its complementary yang partner in Line 1). The aid from Line 1's yang gladdens the yin of Line 4. This is appropriate.

Line 3 It's natural that things come in twos. Life's polar forces seek their own balance. In relationships, generally two is company, three's a crowd. The 3 in itself has a characteristic dynamic tension that forces change. When three form a group, for instance, often the balance of energy is unsettled, and so eventually one will depart. But when one walks alone, one eventually attracts a complementary companion.

Line 2 Stick with what is right and appropriate. The best help is simply to maintain a proper attitude, because taking action can be difficult or offensive. So help passively. Doing this much achieves the right attitude and improves the situation without personal loss.

Line 1 It is right to leave one's own situation and hurry off to help another. But nevertheless ask, "What is this it really for?" Is it a response to another's need, or is it mostly for one's own self-gratification? Giving should cease when it begins to do more harm than good.

Associations with *Starting Small*

Low-Keyed Beginning	*Give to the Cause*	*Decrease for Gain*
Quit Fretting, Focus, Start!	*Voluntary Self-Tax*	*Modest Launch*
Buckle Down and Begin	*Sacrifice Brings Favor*	*Foresighted Effort*

Your Own Association Regarding this Archetype

The Analogy

In Chinese thought, the trigrams of Lake and Mountain suggest two extremes. The mountain is very high; the lake is very deep. They can only interact by decreasing their difference. This happens when moisture from the lake cycles up into the air as cloud, making rain and snow fall onto the dry mountain, washing rocks into mountain streams that travel down into the lake again. All the soil on earth has come from the weathering of rock which enabled plants to grow and so create humus, which then encourages more plant life, and so on. This interchange of energy between the extremes of the soft, low, wet lake and the hard, high, dry mountain starts a creative dynamic going.

Analysis

In *Starting Small*, what seems like a reduction brings eventual gain. Giving from the store of one's own energy, emotion, money, or security plants seed for the future. Temporarily it decreases the stored hoard, true, but eventually this very act ensures the harvest.

It is important to notice that it all begins with something so small as the invisible moisture rising to become cloud on the peak. Something begun sincerely, even though it starts small, brings great rewards. By investing in the future, what seems to be a lessening becomes more. Giving from one's reservoir (physical or mental), if done sincerely, brings great good fortune without error. Right behavior and consistent advantage come from it. How to begin? Even two small containers of seed are enough to make a start, said King Wen. The two main points in this hexagram are first, decrease the random motion and hoarded reserve; second, begin—even though it means starting small. Too often, I believe, this hexagram has been interpreted as a simplistic decrease or decline without explaining the more complex image that is involved here. It is *purposeful* decrease for long-term gain.

Summary: Decrease stored reserves to begin something sincerely, even though it means starting small. Begin in a small way for eventual gain. It brings right action, long-term advantage, and great reward. The two essential points are, first, to focus the stored and undirected energy purposefully, and second, be willing and ready to start off small.

A Modern Analogy for *Starting Small*

Someone once consulted me regarding a problem in his professional life. He was being strongly criticized by his superiors. They wanted him to make some changes in his attitude and behavior or leave.

This man was not sure whether he could make the changes they asked for, or even if he wanted to. He felt personally attacked and defensive about this. Yet he also reluctantly admitted that they just might be right. Perhaps he needed to take a fresh look at himself.

Inner conflict drove him into a quandary. Should he just abandon the professional situation and all his bright hopes for it? Or should he instead develop a new attitude and approach? Maybe. But how, for god's sake? He felt so entrenched in his old habits and ways.

The I Ching's recommendation was Hexagram 41, *Starting Small*. We discussed his associations with this answer, and he decided to initiate a change in himself, not in the outer physical location. He felt he had long carried a counterproductive attitude around, and that he would just perpetuate it by abandoning yet another promising position. So he courageously opted for real change within. It meant giving up his old posture of glib opportunism for a new reliability and integrity.

The result was success. That first year was hard, but his superiors knew he'd made the commitment to change and they admired his grit. He even gained a new self-esteem that was based on reality, not a slick facade. This translated into more success and satisfaction in his work.

Experience has shown me that Hexagram 41 says to stop fretting and wasting your unfocused energy, initiate a small sincere offering, and just begin! Sometimes it is hard, though, to decrease the present comfort zone or habit or fear for a long-term benefit. It may mean sacrificing something intangible—like an entrenched attitude, a passionate attachment, a dominant position, a compulsive urge, a habitual anxiety. Or it may mean foregoing something as concrete as a tempting drink, a drug, a dessert, or a couch-potato session.

Sometimes it seems so hard to initiate a change for the better. The extremes look too polarized—the mountain is too high and the lake too deep. For instance, addiction can trap us into the binary shunt of denial vs. indulgence. It may feel impossible to begin a new path beyond a double bind. How to begin? Even two small containers of seed, said King Wen, are enough to make a start. (The containers can be translated as baskets, bowls or even pods). And *two* containers suggest the male sperm and female egg that together create a child. They remind us that two poles must interact to commence activity. Once there is fluctuating interchange between differences, new circumstances are aborning. Commingling yang action and yin receptivity will rebirth the ego into a new way. Remember, this is the very hexagram that signals Met-Start, the exploded RNA hexagram of AUG.

The Dynamic of Free Will

Our habits can imprison us. Sure, they stabilize and protect us, but they can also recycle mindlessly and chain us into bonds of iterating sameness and inertia. The thrusting line joins the reiterating cycle to bring about a safer spiral of change. Uniting circular habit with linear analysis will trigger the analinear dynamic of evolving alternatives. It is choice that gives us free will.

More insight means more choices. When the psyche recognizes an archetypal pattern that it is mindlessly locked into, it can start to integrate and transcend it. No longer a puppet to knee-jerk anger, envy, greed, it can relax into the free choices opened up by dialogue, relationship, insight. Each new option opens up more free will.

Free will. You alter the old imprisoning patterns. You shift the habitual dynamics even as you start to choose. By working with both mind and heart to harmonize your life, you can alter the pattern of your habits with people, with the global environment, with god. Just by perceiving your life differently, you can begin to live it differently. More consciousness offers more choices.

The psyche evolves so much more swiftly than does the body. Your body is set into hard flesh, but your psyche is a fluid envelope in constant flow. Its cloudy force clusters and hovers and constellates into visibility through shifting events that reveal what patterns you have entrained in the resonant nesting cycles of pattern. By awareness, your psyche can actually evolve its own patterns of energy.

What direction are your habits going? Where do they take your soul? Do you want to go there, or somewhere else? Asking these questions will trigger your free will into seeking the Tao.

But finding the right way and place to start may seem a mystery. How does one enter the untrod way on shod feet? In starting to walk the way of the Tao, you don't even need to know how, but only that you want to. Any point is the point of entry. All places are the place to begin, all moments the time to start.

When I approach the hidden realm, I look to see how parts fit into half-hidden pattern. I become process-oriented, not product-driven. I pause to ponder the resonances among things happening rather than rush for the goal of some final sum. I relax into savoring the joy of being, not just single-mindedly doing. I explore the multitudinous ways that events can shift, how their proportions can flow and alter the whole design. I seek to find an alternative that will enhance the design. I try to face the shadow that I don't even like to admit exists. Every pet hatred of mine will hold a mirror up, I remind myself, to some shadowy aspect that is within me. Only the shadow knows. Out of its dark disorder comes the glint of truth.

A small sincere offering of your desire opens the way. Then the universe resonates in sympathy and starts an appropriate change in you, in your surroundings. You see more, open your eyes to new choices. It means more free will. And you gain it merely by increasing your awareness of hidden pattern and how it operates in you. Increasing your consciousness modifies the analog repetitions. *Educating your ego to see your analog patterns automatically changes them.*

Your ego has a hard task, though, for walking the way of the Tao means befriending the darkness rather than fighting it blindly. When you fight your shadow or deny its existence, you only split it off. Fighting it only polarizes the opposition, but relationship will turn it friendly and benign. Taming your dark side will transform it into a rich dragon-horse of the creative source, befriends your evil enemy into rich mystery containing the gold of unexpected truth.

To appreciate the deep richness of shadow, consider a beautiful set of ornate old silverware that has become tarnished with time. Sure, I could dip each piece into liquid polish and remove all the tarnish. Temporarily. But if I polish it instead with a cloth, the effect will be much more beautiful. It leaves shadow in the incised depths and brings out the luster to higher relief. The whole design becomes evident when we appreciate both dark and light. Allow both poles to exist together. They must both exist somewhere, either together or alienated. If they are closer, they are more friendly and supportive.

If you do not acknowledge your shadow along with light, you lose depth and perspective. You become shallow, lacking in resonance and meaning. Sweetness and light, you'll insist on that. Yet you'll fear the first touch of tarnish. Such denial begs to be slapped by the dark side of nature. It simply sets the bait for the ego trap.

Most of us have polished up our egos pretty thoroughly, so that we can admire ourselves in their bright, shiny surface. We ignore the pits and dull spots. But to integrate my shadow, I must admit and examine and reconcile aspects that my ego doesn't want to notice. I must even consider a hidden, unnoticed pattern to these pits and flaws. What design do they reveal about me? How can I turn this pattern into something more attractive and benign? Don't use guilt, though, but love and enlightened self-interest as the motivators.

A guilt trip does not go to the Tao. It is a self-indulgent wallow in shadow. Do not enter shadow and just stay there, succumbing to its depths and indulging yourself in the misery or shock of how terrible you are. You have turned your shadow inside out and fallen in love with it. This is not analinear balance. It is not the way of the Tao.

That's the trouble with shadow—it doesn't allow us either to bury it or to worship it with impunity. It holds the secret bodies that return as zombies, the undead, the vampires who suck our life blood, the witches who enchant us, the werewolves who go crazy under the full moon. This rife symbolism of horror reveals shadow that is unredeemed, come back to haunt and mesmerize us. Shadow will out. Split off and unredeemed, it will turn alienated and monstrous.

A fellow once told me fiercely, "I'm so angry I could kick a small animal to death." Startled, I gazed at him, thinking, "Yes, but at least you're saying it, not doing it." He went on to discuss what was bothering him. But if he'd repressed that anger long enough, distorted it bad enough, denied it hard enough, he could have kicked viciously at his wife or kid or cat or me, an interested bystander. Repressing the unacceptable into the unconscious and disowning it does not redeem it. Passionately disowned feelings just return another way. It works much like a hydraulic lift—push that force down here and it pops up over there, but it's now much stronger and seemingly unrelated to the casual eye. Beat the wife, hit the child, screw the neighbor, nag at the employee, sabotage the boss, undercut yourself.

But shadow can be dealt with, transformed, befriended, so that paradoxically, it enriches our lives. Medieval alchemy said, "Out of the shit comes the gold." That same old shit keeps recycling until we can manage, with enough awareness, to refine its gold of truth. Wonderfully, people even do it frequently. How often have you heard or said, "It was a terrible experience, but I wouldn't take anything for it now." "Looking back, that failure became the spur to my success." "Mighty Ohs! from little achings grow."

To transform the shadow, you must walk into the analog realm of hidden design. Do you dare to explore one of the esoteric arts? I recommend the I Ching for its unique voice, but dream interpretation, Tarot, astrology, palm reading—all these weird, outrageous and misshapen arts can hold an eerie kernel of truth. We have just distorted them into monsters by misuse and superstition and power trips with the linear whips of our long denial and derision.

Truth waits wherever holistic connection meets with linear logic. Admittedly, the analog domain has become alienated, warped into witchery and animal sacrifice, into the evil eye, black magic and voodoo. But truth still exists behind that monster mask. Friendship will tame and harness it. Enter the hidden hologram and experience its deep beauty. Discover the difference between chance and superstition and synchronicity, between projection and pattern and process that is truly embedded in the design. Learn to see yourself clearly in the mirror of your own fears and hopes. It will heal and change you.

Find your own way into the analog realm and take up a conscious relationship with it. Perhaps it will mean a decision to record your dreams and ponder them, work with them. Do not use the old reductionist view that fears the id as a tar pit of doom. Choose a more transpersonal school like Jungian, Grofian, gestalt, or analinear psychology. Find joy and beauty and comfort in your depths.

Or perhaps your avenue into the unconscious will be the I Ching. Derive a daily hexagram to ponder. Insight will slowly reveal your own inner landscape. The I Ching offers an independent voice of wisdom to dialogue with your own ego. Learn to talk together, and learn to listen. In events, watch for what synchronicity is telling you. Monitor cues and clues semaphoring at the edge of attention. Learn to distinguish the small still voice of truth from the whispers of habit. Recognize the impetus for change that is calling past your projected fears and rages and self-doubt. Establish conscious connection with the greatest teacher, that silent truth beyond words hidden in the dark, waiting to be recognized and loved and attended.

With higher truth, you can create your life anew. You will find the task that you were born for. You will move past your karma (the heavy burden of ritualized past) into your dharma (the transcendent service that you are here to provide the world). Through service, you can grow your soul into a unique crystal in the diadem of life.

The way to free will is the analinear path, melding both analog and linear sides together. This is the opposite of the forking road. Instead you take the merging road to ultimate union with the divine.

All analogs lead home. Weirdly enough, we can even put it into math. Recall Marilyn and Elvis: "Marilyn was to the 1950s what Elvis was to the 1960s." The four terms are A is to B as C is to D. If I say, "Marilyn was to the 1950s what Elvis was to . . . ," then you wait for that fourth term of D. You are *de facto* entrained to expect the rest of this equation, the D that is its punch line. It is archetypal, this relational force. Everywhere, in numbers, habits, even in jokes. It generates the connective tissue of the cosmos.

So how do Marilyn and Elvis and this ratio apply to you and shadow? Well, shadow hides in the hanging, unfinished term. For instance, if a pair in relationship is cast away on the proverbial desert isle—or even just psychologically, like in ancient Greece for Oedipus or Electra, or in a bad marriage, or in a work vendetta— then as a third term is born (the inevitable child of that pair in relationship), it must find its own proper mate. A fourth term must then complete the analog ratio of "Daddy A is to Mommy B as Junior C is to . . . ? Hmm, to Mommy, if I can kill Dad? To a girl like the girl that married dear old Dad?" The fourth term that fills in this archetypal ratio all depends on what energy is constellating around it.

That third term, the child born of the pair's relationship, can kill off one parent and couple with the other. But it ruins the ratio. It destroys the balance of paired pairs in proportional relationship. It is taboo and also bad math. From this very act comes the primal sin of Oedipus. He threw the equation out of whack by reducing the terms back to two again instead of fulfilling it with a new fourth term. Dramas both literary and real have hinged on this fourth term finding its natural or unnatural expression.

This works with everything too, not just sex. Take greed. Abel is to his birthright as I am to . . . ? His birthright? My own? Why not steal my brother's pottage to complete that fourth term? I'll don woolly sheepskin on my arms to fool blind old Dad. Abel is to Bowl as Cain is to Bowl. Then Abel equals Cain and winner take all. Then the universe steps in and says, "Untrue. Bad math. Go back and do the exercise again with a new problem of the same kind."

Or what about this one: "Midas is to his money as I am to . . . ? His money? My own? His, if I can get it?" The ego strives to finish out that ratio with whatever materials are at hand. That is why opening up more options is so very important. That fourth term can actually be finished in a multitude of ways, depending on the options that you can see. What you will see is what you get. So see more.

The hunger to complete some nagging analog ratio is constantly iterated in the body, in the mind, in the universe itself. Once you've got a pair in relationship, then a result springs from it so you've got three terms going, and then everything else is in the pipeline. As Maria Prophetissa put it, as the Tao Te Ching says, as the folk tales all know, that third term foreshadows the ten thousand things. Period 3 implies patterned chaos. Third time's the charm. The 3 *must* initiate change by attracting its fourth term.

This is qualitative math. Proportional, relational. It exists in all things, but it is particularly showcased in art—in music, literature, painting, theater, sculpture, architecture, dance and so on. Take Wallace Stevens' poem on page 191, for instance. It first portrays cycling analog resonance as that angel lazily flies around the abstract garden of Stanza 1. Then the mystery of actual birth into matter starts to gestate within those cattle skulls, aided by the hooded drummers of Stanza 2. The dark drummers are rumbling life from those skulls shaped like the hidden female reproductive organs. Ancient societies often used cow skulls with their secret silhouette of horns and bone to evoke the shape of the uterus and two ovaries. It precedes the babies of Stanza 3, those infants who again go round and round, but now in differentiated and merely human families, born into matter. You are the fourth term connected to it, filling out the ratio of progression into matter with your own real life. Art allows this; it invites and promotes your participation in its resonance. All art holds just such subliminal bonds and resonances. I picked as our example this poem by Wallace Stevens because poetry is relatively easy to print and discuss in microcosm. But this holds everywhere.

Finding that right fourth term is always the tough task. It is the mysterious Other that beckons. For example, it is that fourth physical force called gravity which remains intractable after the other three have been united. It is that "fourth dimension" called time that remains so slippery and unknowable. It is that "fourth function" in the psyche (using the lingo of Jungian typology or the Myers-Briggs Type Indicator) that is called the inferior function because it lies farthest from ordinary consciousness. But that fourth term is also your doorway into the hidden realm—and to creation. Walking that path toward recognition and reconciliation with the other can even lead to the subline otherness of god.

Mysteries are inherent in the fourth term, yes, but they also ask to be solved. They yearn for it, fight for it, demand resolution. We can lift the veil and see this mysterious tension as a normal part of bifurcating co-chaos. We know that two poles (whether in bases, I Ching lines, a couple, whatever) will generate a third condition. Perhaps as a codon, a trigram, a child, a setup of three joke details, or three tantalizing episodes that build a fairy tale. The analog domain then seeks completion in a fourth term that is hidden in the mysterious other. As this happens everywhere, all triplets reach across the void into that chasm toward the punch line that suddenly opens up a whole new perspective on what must be joltingly acknowledged. At this next level of polarity, that hidden fourth term transforms into a complementary triplet bonding the double spiral, the hexagram, even the Greek pantheon of six primary gods and goddesses.

Here the vast panorama of co-chaos unites known with other. Here is the Divine Conjunctio. The Greeks symbolized it in their pantheon on Olympus by pairing the six major gods and goddesses: Zeus, Poseidon, Hades/Hera, Hestia, and Demeter. But standing as we do in the modern age, blindered into the narrow mindset of our culture's tawdry and tiresome first three terms, we cannot see that numinous trinity waiting across the void. Instead, we just blur its beckoning outline into some shadowy, mysterious fourth term, a magnetic "something" out there.

We seek the holy grail of completion and call it the strange fourth dimension of time, or the intractable fourth force of physics, or the inferior fourth function of the psyche, or the enigmatic Mary who tardily completes the Christian Trinity and stands for a lost triumvirate of feminine force. That fourth term, in truth, always expands into *three* polarized terms that are acting in concert—as one DNA triplet balances another triplet, as three dimensions of time balance three dimensions of space, as three quarks balance three antiquarks, as a trigram balances a trigram, as the so-called four functions of psyche typology actually reveal themselves to be the paired triplets of a co-chaos system like the 64 hexagrams, even though the MBTI only recognizes 16 type styles, a structure perhaps parallel to what the ancient Chinese called the central core of the I Ching, the "Golden Sixteen" hexagrams. Yes, this co-chaos pattern hides everywhere, reiterating its eternal analinear force.

Logic does not make the world go round, only straight. Straight into walls sometimes, and dead ends. Linear logic alone can propose the silliest realities. Modern society's emphasis on statistics has turned us into ciphers lost in the masses of sums, so that each unit loses the precious quality of its own identity. Let us regain our identities, reclaim ourselves from the stats and totals and bottom lines. Let us embrace the analog and marry it with the linear, so that we become unified inside and out, carrying our multitudinous diversity and flowing it into a single vast pattern.

How odd it is that East and West have acted so much like the right and left hemispheres of some gigantic global brain. Each side has been busy at its own style of functioning, has long ignored or discounted or fought the other's view of reality. But now we are discovering that East and West have different kinds of knowing to offer each other. The disparity that has so long kept us apart can now make us whole. We can bond the mystical with the physical. We can meld mind to matter. We can reunite our global body and soul.

How? Here in the endless unfolding co-chaos paradigm. It is not rigid and dogmatized and a sanctified mystery only for the few. You participate in it daily, even with each breath that must go both in and out, its two poles shifting in endless analog flux until you finally resolve that polarity by dying into something else, some other reality. Until then, your life here remains open-ended. Windows of opportunity keep appearing, revealing the self-similar designs on scales large and small, familiar in form yet ever new in content. You can open that window wider and see more, adjust the mix of patterns and harmonize the whole in your soul.

This concludes what I have wanted to show you of the co-chaos paradigm. Of course, if there's this much to see, then there's more. Endless windows of opportunity. I will keep looking. But its fractal attractors will outlast me. As this book ends, even when I end, it'll just change partners and dance.

WORKSHOP 12

Plan a program combining discussion and experiences. Select book paragraphs to introduce the following topics—or your own.

Ahead of time, plan how to provide I Ching consulting tools for each member who wants this. Take a poll to see what the preferences are. If appropriate, spend some time in the session counting beads or beans, putting them into containers and so on. Make arrangements for choosing and getting I Ching books.

Opening ritual.

❣ It is time to consult the I Ching. Spread around the room with a bit of space for each member. Ask people to close their eyes and focus in on a question that matters personally. Feel the issue in its fullness. Ask to see it more clearly so that you will know how to regard it, live it well, deal with it appropriately. Seek to find your own true nature and become one with your real purpose and meaning. Now write down your I Ching question to get it clear. Next, consult the I Ching in the method you choose. Study your answer in the I Ching book. Read any changing lines you get and also study the changed hexagram. Do not be discouraged if your answer seems difficult or obscure. Do not be shocked if it even suggests to be angry or less giving or more self-protective. The I Ching seeks to balance your life, not exaggerate further your present overdone tendencies. Be patient and open your mind and heart to the diamond value of hard truth. Record your answer.

Pair with another member to discuss your question and answer. Explore each other's issue sensitively and carefully. Do not take over and tell the other person what to do. Instead, let the I Ching suggest new ways of looking at the issue or reinforce your current feelings about it. Express and follow your associations rather than using the scalpel of logic. Finally, seek to home in on the overall tone of the hexagram and carry that dynamic to help you get clear about this issue and deal with it appropriately.

❣ Discussion: What do you think of the I Ching? Of this book? Write and let me know if you like, in care of the publisher. I wish you the best of light and dark, and a life full of deep joy.
Closing ritual. Five minutes of feedback. Announcements.

Bibliography

Anthony, Carol. *A Guide to the I Ching*. Anthony Publishing Company. 1980.

Barrow, John D. *The Anthropic Cosmological Principle*. With Frank Tipler. Oxford: Oxford University Press. 1988.

Benson, Frank, editor. *The Dual Brain: Hemispheric Specialization in Humans*. New York: The Guilford Press. 1985.

Borek, Ernest. *The Code of Life*. New York: Columbia University Press. 1969.

Bowers, Kenneth, & Meichenbaum, Donald. *The Unconscious Reconsidered*. New York: John Wiley & Sons. 1984.

Buchanan, Keith. *China: The History, the Art, and the Science*. With Charles P. FitzGerald & Colin A. Ronan. New York: Crown Publishers. 1981.

Butler, Christopher. *Number Symbolism*. London: Routledge & Kegan Paul. 1970.

Campbell, Joseph. *The Masks of God*. New York: Viking Press. 1964.

Capra, Fritjof. *The Tao of Physics*. New York: Bantam Books, 1977.

Carus, Paul. *Chinese Astrology: Early Chinese Occultism*. La Salle, Ill.: Open Court. 1974.

Casti, John L. *Alternate Realities: Mathematical Models of Nature and Man*. New York: John Wiley & Sons. 1989.

Cole, K.C. *Sympathetic Vibrations: Reflections on Physics as a Way of Life*. New York: Bantam. 1984.

Cook, Norman D. *The Brain Code*. London: Methuen. 1986.

Coward, Harold. *Jung and Eastern Thought*. New York: State University of New York Press. 1985.

Culling, Louis. *The Incredible I Ching*. New York: Samuel Weiser. 1969.

Da Liu. *I Ching Numerology: Plum Blossom Numerology*. New York: Harper & Row. 1950.

Danchin, Antoine & Gascuel, Olivier. "Data Analysis Using a Learning Program, a Case Study: an application of PLAGE to a Biological Sequence Analysis." Paper from the 8th European Conference on Artificial Intelligence (ECAI). 1988.

Davies, Paul. *The Cosmic Blueprint*. London: Unwin Paperbacks. 1989.

Davis, Philip & Hersh, Reuben. *The Mathematical Experience*. Boston: Houghton Mifflin. 1981.

De Bary, William Theodore; Wing-tsit Chan; Watson, Burton. *Sources of Chinese Tradition, Vol. 1 & 2*. New York: Columbia University Press. 1960.

Dewdney, A.K. "Computer Recreations." *Scientific American*, August 1985.

Doczi, György. *The Power of Limits*. Boulder: Shambhala Publications. 1981.

Ellenberger, H.F. *The Discovery of the Unconscious*. New York: Basic Books. 1970.

Fairbank, John. *The Great Chinese Revolution: 1800 to 1985*. New York: Harper & Row. 1986.

Fiedeler, Frank. *Die Wende: Ansatz einer genetischen Anthropologie nach dem System des I-ching*. Berlin: Kristkeitz 1977.

Yin und Yang: *Das kosmische Grundmuster in den Kulturformen Chinas*. Cologne: DuMont, 1993.

French, A.P. *Vibrations and Waves*. New York: Norton, 1971.

Fromm, Erich. *The Forgotten Language; An Introduction to the Understanding of Dreams, Fairy Tales and Myths. New York: Grove Press, 1951.*

Fung Yu-lan. *A History of Chinese Philosophy*. Trans. Derk Bodde. Princeton: Princeton University Press. 1952.

Gao Heng. Article translated from *Wenshizhe* by Edward L. Shaughnessy. *Zhouyi Network Newsletter*, No. 1, Bowdoin College, Brunswick, Maine, January 1986.

Gardner, Howard. *Art, Mind and Brain: A Cognitive Approach to Creativity*. New York: Basic Books, Inc. 1982.

Gardner, Martin. "Mathematical Games" in *Scientific American*, January, 1974.

Gatlin, Lila. *Information Theory and the Living System*. New York: Columbia University Press. 1972.

Glass, Leon & Mackey, Michael. *From Clocks to Chaos: the Rhythms of Life*. Princeton: Princeton University Press. 1988.

Gleick, James. *Chaos: Making a New Science*. New York: Viking. 1987.

Gordon, Rosemary. "Reflections on Jung's Concept of Synchronicity," *Harvest*, Vol. 8, pp. 77-98. 1962.

Heisenberg, Werner. "Scientific and Religious Truths." Seen as a typed essay.

Hofstadter, Douglas. *Metamatical Themas: Questing for the Essence of Mind and Pattern*. New York: Basic Books, Inc. 1985.

Hook, Diana Ffarington. *The I Ching and Mankind*. London: Routledge & Kegan Paul. 1971.

Hughes, E. R., trans. & editor. *Chinese Philosophy in Classical Times*. New York: E.P. Dutton & Co. 1942.

Jaynes, Julian. *Origin of Consciousness in the Breakdown of the Bicameral Mind* Boston: Houghton Mifflin. 1976.

Jenkins, R.C. *The Jesuits in China*. London. 1894.

Jin, Guantao, Fan Dainian, Fan Hongye, & Liu Qingfeng. "The Evolution of Chinese Science and Technology" in *Time, Science, and Society in China and the West*. Volume V of *The Study of Time*. Edited by Fraser, J.T., Lawrence, N., & Haber, F.C. Amherst: University of Massachusetts Press. 1986.

Josey, Alden. "Molecules as Mandalas." Unpublished essay.

Jung, C.G. "Four Lectures on the Chakra Symbolism of Tantric Yoga and the Kundalini System" (1932). New York: *Spring* Annuals of 1975 & 1976.

Jung Young Lee. *Principles of Change: Understanding the I Ching*. Secaucus, New Jersey: 1971.

Kelso, J.A.S., Mandell, A.J., & Shlesinger, M.F. *Dynamic Patterns in Complex Systems— Conference Proceedings*. Teaneck, New Jersey: World Scientific. 1988.

Kramer, Johnathan D. "Temporal Linearity and Nonlinearity in Music." in *Time, Science, and Society in China and the West; The Study of Time, Vol. V.* J. T. Fraser, N. Lawrence, & F. C. Haber, editors. Amherst: University of Massachusetts Press. 1986.

Kreutzer, Carolin S. "Archetypes, Synchronicity and the Theory of Formative Causation," *Journal of Analytical Psychology*, Vol. 27, pp. 255-262. 1982.

Kuhn, Thomas. *The Copernican Revolution*. Cambridge, Maine: Harvard University Press. 1957.

Lattimore, Owen. *Studies in Asian Frontier History*. London. 1962.

Lepore, Franco; Ptito, Maurice; Jasper, Herbert, editors. *Two Hemispheres—One Brain: Functions of the Corpus Callosum*. New York: Alan R. Liss, Inc. 1984.

Legge, James, trans. *I Ching: Book of Changes*. Published in English in 1899.

Leibniz, G. Wilhelm. *Zwei Briefe über das Binare Zahlensystem und die chinesische Philosophe*. Munich: Belser. 1968.

Lex, Barbara. See D'Aquili, Eugene G., Laughlin, C.D., & McManus, J. *The Spectrum of Ritual: A Biogenetic Structural Analysis*.

Liu Dajan. *A Preliminary Investigation of the Silk Manuscript Yjing*. Trans. from *Wenshizhe* by Edward L. Shaughnessy. Shandong: Shandong University. 1985.

Lovelock, James. *Gaia: A new look at life on earth*. Oxford: Oxford University Press. 1979.

Mandelbrot, Benoit. *The Fractal Geometry of Nature*. New York: Freeman, 1977.

Milner, Brenda, editor. *Hemispheric Specialization and Interaction*. Cambridge, Massachusetts: MIT Press, 1974.

Moore, Steve. *The Trigrams of Han: Inner Structures of the I Ching*. Wellingborough, England: The Aquarian Press. 1989.

Morris, Eleanor B. *Functions and Models of Modern Biochemistry in the I Ching*. Taipei: Cheng Chung Book Company, 1978,

Nalimov, V.V. *Realms of the Unconscious: The Enchanted Frontier*. Philadelphia, Pa.: ISI Press. 1982.

Needham, Joseph.
   Chinese Astronomy and the Jesuit Mission: an Encounter of Cultures. The China Society. 1958
   "Science and China's Influence on the World," in *The Legacy of China*, edited by Raymond Dawson. Oxford: Oxford University Press. 1964.

Newmark, Joseph; Lake, Frances. *Mathematics as a Second Language*. New York: Addison-Wesley. 1982.

Newsweek. "An Instinct for Survival: Chinese students and professors await a new day." July 24, 1989.

Ni, Hua Ching. *The Book of Changes and the Unchanging Truth*. Los Angeles: College of Tao & Traditional Chinese Healing. 1983.

Pagels, Heinz. *Perfect Symmetry*. New York: Simon & Schuster. 1983.

Peat, F. David. *Synchronicity: The Bridge Between Mind and Matter*. New York: Bantam. 1987.

Peitgen, H.-O., & Richter, P.H. *The Beauty of Fractals*. Berlin: Springer-Verlag. 1986.

Prigogine, Ilya. *From Being to Becoming: Time and Complexity in the Physical Sciences*. San Francisco: W.H. Freeman & Company. 1980
   Order Out of Chaos: Man's New Dialogue With Nature. With Isabelle Stengers. New York: Bantam Books. 1984.

Rothenberg, Albert. "The Emerging Goddess." *The Creative Process in Art, Science, and Other Fields*. Chicago: The University of Chicago Press. 1979.

Salisbury, Harrison. "A View From Mount Lu: Shedding 'A Little Blood.'" Paris: *International Herald Tribune*. June 15, 1989.

Schipper, Kristofer. Wang Hsiu-huei. "Progressive and Regressive Time Cycles in Taoist Ritual" in *Time, Science, and Society in China and the West; The Study of Time, Vol. V*. J. T. Fraser, N. Lawrence, & F. C. Haber, editors. Amherst: University of Massachusetts Press. 1986.

Schoenholtz, Larry. *New Directions in the I Ching*. Secaucus, New Jersey: University Books. 1975.

Schönberger, Martin. *The I Ching and the Genetic Code: the Hidden Key to Life*. New York: ASI Publishers. 1979.

Sheldrake, Rupert. *A New Science of Life*. Los Angeles: Tarcher. 1981.

Shubnikov, A.V., & Koptsik, V.A.; trans. editor, Harker, David. *Symmetry in Science and Art*. New York: Plenum Press, 1974.

Stent, Gunther. *The Coming of the Golden Age*. New York. Natural History Press. 1969. *The Double Helix*, ed. by Gunther Stent. Norton Critical Edition. New York: W.W. Norton & Co. 1980.

Stevens, Wallace. *The Collected Poems of Wallace Stevens*. London: Faber and Faber Limited. 1955.

Sung, Z D. *The Symbols of Yi King*. Shanghai: The China Modern Education Company. 1934.

Suzuki, David; Griffiths, Anthony; Miller, Jeffrey; Lewontin, Richard. *An Introduction to Genetic Analysis*. New York: W.H.Freeman. 1986.

Swami Ajaya. *Psychotherapy East and West: A Unifying Paradigm*. Honesdale, Penn: Himalayan Institute of Yoga Science. 1983.

Tien-Yien Li and Yorke, James A. "Period Three Implies Chaos." *American Mathematical Monthly*, Vol. 82. December, 1975.

Toffler, Alvin. *Future Shock*. New York: Bantam. 1971.

Vasavada, Arwind. *Hinduism and Jungian Psychology*. With Spiegelman, J. Marvin. Phoenix, Arizona: Falcon Press. 1987

Von Franz, Marie-Louise. "Dialog über den Menschen." Stuttgart: Klett Verlag. 1968. *Number and Time*. Evanston: Northwestern University Press. 1974.

Waley, Arthur. *The Book of Songs*. Boston: Houghton Mifflin. 1937. *Three Ways of Thought in Ancient China*. London: Allen & Unwin. 1939.

Watson, James. *Molecular Biology of the Gene*. New York: Benjamin, Inc. 1965.

Whincup, Greg. *Rediscovering the I Ching*. Garden City, N.Y.: Doubleday. 1986.

Wiggins, Stephen. *Global Bifurcations and Chaos: Analytical Methods*. New York-Berlin: Springer-Verlag. 1988

Wilhelm, Richard, trans.Cary F. Baynes. *The I Ching: Book of Changes*. Princeton: Princeton University Press. 1950.

Wilhelm, Hellmut. *Change: Eight Lectures on the I Ching*. Trans. C.F. Baynes. Princeton: Princeton University Press. 1972.

Yan, Johnson F. *DNA and the I Ching: The Tao of Life*. Berkely, CA: North Atlantic Books. 1991.

Acknowledgments

We gratefully appreciate permission to quote brief passages in the following works:

Page 13, four lines are quoted from the poem "Auguries of Innocence" by Robert Burns, written 1800-1808.

Page 38, Philip Davis and Reuben Hersh, from *The Mathematical Experience*. Boston: Houghton Mifflin. 1981.

Page 46, Joseph Needham, from "Science and China's Influence on the World," in *The Legacy of China*, edited by Raymond Dawson. Oxford: Oxford University Press. 1964.

Page 47, Owen Lattimore, from *Studies in Asian Frontier History*. London. 1962.

Page 50, John Fairbank, from *The Great Chinese Revolution: 1800 to 1985*. New York: Harper & Row. 1986.

Page 58, Leon Glass and Michael Mackey, from *From Clocks to Chaos: the Rhythms of Life*. Princeton: Princeton University Press. 1988.

Page 64, Kristofer Schipper and Wang Hsiu-huei, from "Progressive and Regressive Time Cycles in Taoist Ritual" in *Time, Science, and Society in China and the West; The Study of Time V*. J. T. Fraser, N. Lawrence, & F. C. Haber, editors. Amherst: University of Massachusetts Press. 1986.

Page 79 Benoit Mandelbrot, from *The Fractal Geometry of Nature*. New York: Freeman, 1977.

Richard Wilhelm is quoted several times in Cary F. Baynes translation of *The I Ching: Book of Changes*. Princeton: Princeton University Press. 1950.

James Legge's classic translation of the I Ching, first published in England in 1899, is quoted various times, but especially in the chapter called "The Master Plan."

Page 83, A.K. Dewdny, from "Computer Recreations." *Scientific American*, August 1985.

Pages 86 and 107, James Gleick, from *Chaos: Making a New Science*, copyright by JAmes Gleick in 1987. Published in New York by Viking in 1987.

Page 87, H-O Peitgen and P.H. Richter, from *The Beauty of Fractals*. Berlin: Springer-Verlag in 1986.

Page 91, Steven Weinberg, from *The First Three Minutes (updated): A Modern View of the Origin of the Universe*. New York: Basic Books, 1988, now with HarperCollins.

Page 101, Paul Davies, from *The Cosmic Blueprint*. London: Unwin. 1989.

Pages 132 and 147, Douglas Hofstadter in *Metamatical Themas: Questing for the Essence of Mind and Pattern*. Copyright by Basic Books in 1985, now with HarperCollins.

Page 133, Lila Gatlin, from *Information Theory and the Living System*, Columbia University Press. New York. 1972.

Page 134, Werner Heisenberg, from "Scientific and Religious Truths." Seen as a typed essay.

Page 137, Thomas Berry, from *The Dream of the Earth*, published by Sierra Club Books, San Francisco in 1988.

Page 139, Rupert Sheldrake, from *A New Science of Life*. Los Angeles: Tarcher. 1981.

Page 191, "The Pleasures of Merely Circulating" by Wallace Stevens is quoted from *Collected Poems* by Wallace Stevens, 1936. Permission from Alfred A. Knopf, Inc.

Page 195, Jonathan Kramer, from his essay in *Time, Science, and Society in China and the West; The Study of Time V*. J. T. Fraser, N. Lawrence, & F. C. Haber, editors. Amherst: University of Massachusetts Press. 1986.

Several Chinese authors were quoted who proved impossible to contact. Nevertheless, their scholarship is gratefully appreciated.

Illustration Acknowledgments

The vortex illustration on page 59 is adapted from *From Clocks to Chaos: the Rhythms of Life* (page 32) by Leon Glass & Michael Mackey. Princeton University Press, 1988.

The Julia sets on page 86 are images from a 1989 catalog of Art Matrix, Ithaca, N. Y.

The graph illustration on page 96 is adapted from *Fractals, Chaos, Power Laws* (page 280) by Manfred Schroeder. W.H. Freeman and Company Press, 1990.

The Nepalese yantra on page 100 is a computer rendition of a photograph of a yantra in *Yantra: the Tantric Symbol of Cosmic Unity* (page 64) by Madhu Khanna. Thames and Hudson, New York.

The Golden Mean designs on page 206 are collated and adapted from *The Power of Limits* by György Doczi, Shambhala Publications, Boulder, Colorado, 1981.

The molecular diagram which comprises the upper left portion of the illustration on page 239 is a notebook adaptation of a diagram on page 132 of *Molecular Biology of the Gene* by James Watson, published in 1965 by Benjamin Company in New York.

The remaining graphics were done by Katya Walter on several Macintosh computers and a couple of Hewlett-Packard printers. The computer images in the chapter called "Art in Number and Image" were developed using border fonts that were collated and fractally adapted from three computer disks of Decorative Border Fonts, Vols. 45, 46 & 48 by Sunshine, Box 4351, Austin TX 78765.

This book was created using WriteNow and Aldus PageMaker on several Macintosh computers. Camera-ready output was prepared using a HP LaserJet 4M at 600 dpi and then final linotronic printing from disk to metal plates. Graphics were produced using Aldus SuperPaint, Adobe Photoshop, and Adobe Illustrator. As for the text, it is mostly set in Palazzo typeface, a friendlier variant of the familiar Palatino, while the graphics mostly use the wide Chantilly family of fonts. These *definiType* fonts come from SoftMaker, 2195 Faraday Ave. Suite A, Carlsbad CA 92008-7217.

About the Author

Katya Walter received a Ph.D. with an interdisciplinary emphasis from the University of Texas in Austin. She spent five years of post-doctoral study at the Jung Institute in Zurich. She has taught in universities in the United States and China, and has published in the areas of cosmology, social analysis, fiction, and poetry. In Austin, Texas, she sees clients in analinear analysis, lectures, does workshops, and writes on life as an expression of chaos patterning, with an emphasis on spiritual growth. She is married and has two grown children, two dogs and two cats. She maybe reached by Internet email at kairos@uts.cc.utexas.edu

About the Kairos Center

The Kairos Center was founded to explore the integration of body and soul. It is centrist and yet it investigates the edges of spiritual reality. It honors the best of old and new, left and right, East and West, physical and spiritual, linear and analog domains. Our members work for love as well as money, and our network taps a wide range of expertise from theory lecturers to experiential facilitators to body workers and Chinese medicine practitioners.

Ordering Information

If you want to order fractal slides or a video showing the changing fractal jewels of the Mandelbrot set, contact the people who made the fractal for this cover: Art Matrix, Box 880, Ithaca NY 14851-0880 USA. Phone: 1-800-PAX-DUTY. FAX: 1-607-277-8913.

If you want to order a set of I Ching stones in a fabric pouch, call for current pricing. If you want to be on the Kairos Center mailing list for news about workshops, classes, or spiritual tools, call, write or FAX to:

Kairos Center

Box 26675
Austin, Texas 78755
Phone 1-800-624-4697
FAX: 512-453-8378